BRINGING BACK THE DODO

LESSONS IN NATURAL AND UNNATURAL HISTORY

WAYNE GRADY

McCLELLAND & STEWART

Library and Archives Canada Cataloguing in Publication

Grady, Wayne
    Bringing back the dodo : lessons in natural and unnatural history / Wayne Grady.

ISBN: 978-0-7710-3504-3 (bound)
ISBN: 978-0-7710-3505-0 (pbk)

    1. Nature — Effect of human beings on. 2. Evolution (Biology)
3. Human ecology. I. Title.

GF75.G69 2006   304.2   C2005-907320-9

We acknowledge the financial support of the Government of Canada through the Book Publishing Industry Development Program and that of the Government of Ontario through the Ontario Media Development Corporation's Ontario Book Initiative. We further acknowledge the support of the Canada Council for the Arts and the Ontario Arts Council for our publishing program.

Typeset in Bembo by M&S, Toronto
Printed and bound in Canada

McClelland & Stewart Ltd.
75 Sherbourne Street
Toronto, Ontario
M5A 2P9
www.mcclelland.com

1 2 3 4 5   11 10 09 08 07

"Nothing spreads faster than Science, when rightly and generally cultivated."

– John Dryden

"Alas! Can we ring the bells backward? Can we unlearn the arts that pretend to civilize and then burn the world? There is a march of Science; but who shall beat the drums for its retreat?"

– Charles Lamb

"If a vote could undo all the technological advances of the last three hundred years, many of us would cast that vote, in order to safeguard the survival of the human race while we remain in our present state of social and moral backwardness."

– Arnold Toynbee

Contents

FIRST WORDS

Five years ago, James Little, *explore* magazine's intrepid editor, called and asked me to write a regular column for his publication. What he had in mind, he said, was for me to be "the Canadian David Quammen," the American naturalist who for many years wrote a natural-history column for *Outside* magazine. I was a bit taken aback by this, as I'd always thought of myself as the Canadian John McPhee, but adaptability is life, and so I took it on. I'm very glad I did, for writing the column got me thinking about human beings as a species in ways I hadn't thought of before. The idea was to provide a natural history slant on contemporary life; every two months, I was to look at what was happening in the world around us from the point of view of an amateur naturalist. The column was called "Biologic," and I eventually wrote fifteen of them, each about twenty-five hundred words in length. Long for a column, but it was good to have enough room to follow various unravelled threads and to end up, one hoped, with a tidy skein of thought. James gave me free range to explore as wide a swath of scientific, natural, and human foibles as I chose. He suggested general ideas for some of the columns, and readers wrote or e-mailed in with anecdotes that sparked others, and over the two and a half years of Biologic's life the columns evolved into a unified, though eclectic, collection of essays.

Well, nascent essays. Although there is a school of thought that maintains the publication of a collection of columns ought to be simple reprintings of the original columns, presenting them exactly as they were published, warts, dated references, factual errors and all, I do not attend that school. Columns are not essays. They are subject to specific, often passing, fancies; they are written to firm, often pressing, deadlines; and they are squeezed into iron-clad, often restrictive, lengths. Essays require a somewhat more relaxed environment in which to thrive. In this book, like a conscientious gardener, I have dug the columns from their temporary seed beds – the pages of *explore* – and transplanted them into more permanent perennial beds. In many cases I have added information, updated time-sensitive references, written new paragraphs, and, in two cases, entire essays ("Atwood and McKibben" is an expansion of a review I wrote for the *Ottawa Citizen* of Margaret Atwood's novel *Oryx and Crake*, and "Send in the Clones" dovetails some of the ideas expressed in the previous essays). I have corrected factual errors wherever I (or others) have found them, expanded several of the essays beyond their original scope, deepened them, and generally tried to make them reflect the maturer thoughts of a writer who has had the time and tranquility to mull over the ideas worried at in the columns. The result, I hope, is a book of connected essays with a unified theme and a single, sustained voice.

But what is that unified theme? I am not being coy or evasive by saying that if I could express it here in a few words I would not have had to write the essays. But like any reader, I have had to go over these pages myself in order to discern what themes emerged. Generally speaking, in these essays I seem to be constantly alarmed at our tendency to ignore or deny the degree to which we are part of the natural world. I believe it is true that, as J.F. Blumenbach, the nineteenth-century founder of anthropology, first observed, we are "the most perfect of all domesticated species." Many of these essays are ruminations about what that means. But we have not taken nature out of ourselves – even the most domesti-cated cat eats, drinks, breathes, hunts, hosts fleas, and repro-duces – rather, we have taken ourselves out of nature. To our cost. In many of the essays I try to remind us of the fact that when we destroy a segment of nature – by cutting down a forest to make a road, or killing wild animals for sport, or even ridding ourselves of pests and parasites – we destroy an essential part of ourselves. When we tamper with nature, by altering an organism's genetic makeup to produce a new plant or animal, or bypass sexual reproduction through cloning or gene splicing, when we remove a species from or add a species to an ecosystem, we are interfering with a process that has evolved on its own, and which has taken us into account, for millions of years, and about which we know next to nothing. It ought to be a sobering thought that, when most of us encounter a bear in the forest, the bear knows more about us than we know about it.

I am not, however, a polemicist by nature. My inclination is simply to point out what we're doing as a species, place that action in some kind of natural context, and occasionally ask why we persist in doing it. If the voice sometimes sounds plaintive, or incredulous, or impatient, well, that is often the voice of the essayist. An essay is a pearl that began with an irritating grain of sand.

Kumquats from Colombia

Of all the eclipsed names in natural history, that of Matous Klácel stands out, if that's the right phrase, for the particular density of the obscurity in which he worked. Klácel was an Augustinian monk who tended a tiny garden in a remote monastery called St. Thomas's, in Brünn, a small town in Moravia, in the 1840s. He called his patch his "alpine garden," because in it he transplanted various flowers and shrubs he collected from the nearby mountains; he wanted to see how upland plants would do in the lowlands, where the monastery was located. He maintained his experimental garden for more than a decade, making meticulous observations of each plant species through a dozen generations. He had asked himself a very serious question, but the conclusion he drew from his experiments was so unstartling that it provides a classic example of how negative results can be as important as positive ones, though they are often far less sensational. To his question, Do changes in environment lead to permanent changes in plants?, came the unspectacular answer, No, they do not.

I thought about Klácel the other day as I pushed my shopping cart through the produce section of our local supermarket. As Canadians, we tend to pride ourselves on the degree to which we have adapted to our northern environment. Adaptability is, after all, one of the more desirable Darwinian traits, and there is no doubt that, like Klácel's mountain plants, *Homo sapiens* – a subtropical, grasslands species – seems to be doing pretty well in this mid-temperate to subarctic climate. Whether we have migrated here from Asia or Africa or Europe, most of us are able to find food and procreate in Canada, which is the bare-bones Darwinian definition of success. But have we, in any real sense, *adapted* to our new environment? If Klácel was right, then no species, however hardy, can truly adapt to a new environment in the few generations that most of us have been here.

How, then, to resolve this apparent contradiction between the laws of natural selection and our own success in these northern climes? If we haven't adapted, then how is it we're still here? My tour of the Loblaws produce section suggested one answer: instead of adapting, we've simply recreated our original subtropical climate here in the north. What struck me in Loblaws was that an astonishing number of the fruits on the shelves and in the bins came from the south. Yes, there was a nice assortment of apples, some of which are native to the temperate zones of Europe and Asia. But there was a much greater selection of oranges, which came from Portugal, Spain, South Africa, Israel, and Florida. There were also pineapples from Costa Rica, figs from Greece, papayas and mangos from Brazil, plantain and kumquats from Colombia, star fruit from Taiwan, something called kiwano melons from Ecuador, limes and jicamas from Mexico, coconuts from the Dominican Republic, kiwi (the fruit, not the bird) from New Zealand, medjool dates from Israel, casaba melons, lemons, and tangerines from northern Argentina, honey dates from Iran, Korean pears from China, Asian pears from Chile, Rocha pears from Portugal, and Abati and prickly pears from Italy.

Certainly, we can attribute the presence of this groaning cornucopia of subtropical goods to modern technology, the underlying philosophy of which is: if you *can* have it, you might as well. I must have appeared suspicious as I made a list of all the above goodies, for the produce manager hurried over and asked me if there was something I couldn't find. I was tempted to ask him why there were no lichee nuts in evidence, but instead I assured him I was impressed by the variety of fruit he had on offer, and from so many places around the world. He smiled and said, "Yes, ten or fifteen years ago it would have been oranges and pineapples from Florida and maybe bananas from Dominica, and that would have been it. It's amazing what modern technology has done, isn't it?"

I recalled being on the Caribbean island of Dominica a few years ago, and watching men loading bananas from a seemingly endless line of ancient trucks into a huge cargo ship. Nowadays they would be stuffing them into the belly of a large aircraft. Technology accounts in part for *how* the produce gets here. But as nature writer Barry Lopez has argued, it isn't the airplane alone to which we owe our imported largesse – after all, we've had airplanes up here a lot longer than we've had kiwano melons – but rather the marriage of airplanes and computers. Together they are able to guarantee precise arrival times for "perishable" goods, a category that is no longer restricted to food items but now embraces any time-sensitive commodity, such as newspapers, movies, and computer games, even fashionable shades of lipstick. The term "perishable" is now a marketing, not an agricultural, term. If a teenager might consider it passé tomorrow, it's perishable today. Lopez views the Boeing 747 cargo plane as "the ultimate embodiment of what our age stands for," the modern equivalent of the Gothic cathedral.

What it stands for must be hard-won gratification, then, at the expense of good taste or even convenience. To the airplane and the computer, by the way, I would add the many technological interventions, from refrigeration to biochemical growth inhibitors, that allow workers in, say, South Africa, to pick fruit while it is still green and spray it with a hormone-based retardant so that it ripens in a Customs shed outside Montreal and arrives "fresh" at the corner grocer's up to a month later. We can no longer afford to wait for fruit to ripen before harvesting it. A science-minded friend once told me he was driving along a highway in Florida behind a truckload of produce when some tomatoes flew off and landed on the road ahead of him. Instead of receiving a grille-full of gooey spaghetti sauce, as he expected, he watched the tomatoes hit the road, bounce right over his car, land on the road behind him, and keep on bouncing. He stopped to pick one up, and it wasn't even bruised. "Major-league baseballs would have sustained more damage," he said.

If we want the stuff badly enough, someone will figure out a way to get it to us. The real question is *why* we want it so much. Why do we build airplanes with cargo holds twice as long as the distance of the Wright brothers' first flight, which cost $155 million apiece to buy and $15,000 an hour to run, simply so that we can enjoy, as Lopez puts it, "a fresh strawberry on a winter morning in Toronto?"

The answer can be only that, as a subtropical species, we still prefer subtropical foods. Like Klácel's hardy alpine shrubs, transplant ourselves where we will, we do not adapt in any permanent way. Or at least we haven't yet. As Darwin put it, our recent forebears did not pass new traits along to their offspring to make us better adapted to our new environment. To expect such instant adaptation would be an oversimplification of how evolution works. It isn't that direct. Human beings are a twig that branched off from the primate trunk sometime in the mid-Pliocene, about 5 to 7 million years ago, at a time when global climate change was busy turning much of Africa's vast forested regions into grasslands. Over the next several million years (that's how long it takes), by the merest accident of arbitrary gene distribution, one in a thousand or a million of us was born able to cope with the new environment. Eventually, that one met up with another mutant coper and together they had slightly better adapted offspring; over countless generations sufficient offspring with the newly acquired coping trait had been produced to perpetuate the species. We had adapted to our new environment. We became a new species: a grasslands species.

We are still a grasslands species. In the grasslands we decided to walk upright (so we could see where we were going), and it was grasslands food that ensured not merely our survival but also our proliferation. Suddenly we could eat our fill. A single square kilometre of grassland is capable of supporting eighteen thousand kilograms of animals, whereas the same area of forest can sustain barely half that. The principal grassland product is, of course, grass, and we still eat a lot of grass. In fact, such grasses as corn, wheat, and rice remain the bulk foods of our diet. For meat we turned to the principal grassland prey animals, which were grazers, and grazing animals are still our primary source of protein. The only carnivores we eat, as a rule, are fish, a habit we probably picked up hanging around the rivers that flowed through the grasslands. A recent study has suggested that we owe our inconveniently large brains (from the point of view of childbirth) to the crayfish and shrimp that we ate when we first came out of the trees.

As the Pliocene progressed, the average global temperature soared from about 8 to more than 10 degrees Celsius, which means that in the grasslands, which straddled the Equator, the average temperature was about 20 degrees Celsius. This allowed us to expand our territory without moving into new climate zones. We first migrated east and west, not north and south, following the 20-degree Celsius thermal, first into Asia and eventually into the subtropical Americas. This brought us into contact with subtropical forests, especially high-altitude rain and cloud forests. That east-west equatorial belt is where virtually all the food we now eat originally came from.

Although advances in technology (the wheel, the stirrup, the sail) and agriculture eventually allowed us to colonize the temperate zones, rather than adapt to our new environment we industriously set about turning it into an imitation of our natural subtropical habitat. We insulated and heated our buildings to maintain a constant grassland environment – 20 degrees Celsius, 60 per cent humidity, maintained summer and winter, with indoor plants and grassland pets (cats, dogs, small rodents). We cut down the forests to create grasslands, which we called farms, little microsavannahs featuring vast expanses of various imported grasses sprinkled with herds of imported grazing animals.

Where there already were native varieties – on the Prairies, for example – we got rid of them. We replaced bison with cud-chewers from the subtropics. Modern cattle are descended from the earliest domesticated European ox, *Bos brachyceros*, brought north by Neolithic peoples in their wanderings from Asia some seven thousand years ago, along with the earliest pigs and sheep. We ploughed up native northern grasses and planted our own grasses, wheat and corn, which are also descended from more familiar, subtropical species. Wheat, for example, is native to the Middle East, and was domesticated during Neolithic times from a species of wild forage grass known as *Aegilops*. As for corn, there is no doubt that it was a native North American plant, crucial to the Aztec diet two thousand years ago and brought to Europe by Columbus after his foray into central Cuba in 1492. But we have chucked out maize, the original variety, the form in which it was known in Canada before European contact, and plant instead a later type known as dent corn, a hybrid of two varieties from much farther south.

Instead of capturing and taming native meat birds, such as geese, ducks, and grouse, we brought in chickens, bred from, guess what, a more familiar subtropical species, *Gallus gallus*. The only place I've seen wild *Gallus gallus* was on Kauai, one of the smaller Hawaiian islands. It looked exactly like a domestic chicken, only slightly smarter: it actually hurried across the road in front of our car.

In almost every case I can think of, we either ignored or eradicated what existed in the north and substituted something from the tropics or subtropics. Virtually nothing we eat now is native to this part of the world. In his book *Why We Eat What We Eat*, American food writer Raymond Sokolov identifies "American food" as all the plants and animals native to this hemisphere (north of the subtropical zone, including Canada) and unknown to the Old World before Columbus. It's a small if idiosyncratic list: the wild turkey, the American persimmon, black walnuts, sassafras, chili peppers, blueberries, Olympia oysters, wild rice.

I would add a few more neglected items to Sokolov's list. Jerusalem artichokes, for example, were discovered in New England and Nova Scotia by Samuel de Champlain. We don't eat a lot of Jerusalem artichokes; early settlers fed them to their pigs. And if Sokolov can include Olympia oysters, a small bivalve found off the Oregon coast in Puget Sound, then I can include Malpeque oysters, a large bivalve found off the coast of Prince Edward Island, in Northumberland Strait. And if blueberries are on, why not saskatoon berries, a Canadian variety of the European amelanchier, or serviceberry?

If we were truly adapted to our new environment, those would be the most common food items in our pantries. They are not. Native groups harvest and sell wild rice as a specialty food, but no one cultivates it. I have never eaten an American persimmon, and certainly have never seen one on my grocer's shelves. I'm not sure I would recognize sassafras if I tripped over it. True, we eat black walnuts and pick wild blueberries, but neither has replaced imported or introduced nuts and berries such as pecans and almonds or raspberries in our normal diet. We eat turkey maybe twice a year. We find it a dry bird, and look for ways to make it more palatable, moister, more like an ostrich. Food writer James Beard, in the most recent edition of his book *Beard on Birds*, laments the loss of the ruffed grouse, "that champion of all game birds," from North American markets; they are no longer sold not because they are no longer available but because nobody bought them. And hands up all those who have gorged themselves on saskatoon berries lately. Now hands up all those who have eaten a strawberry, trucked or flown up from Mexico or southern California, in the last month. Yes, we go through a lot of chili peppers, but despite their presence on Sokolov's list they are really a subtropical species; they come from the same part of the world as yams, green peppers, and avocados, which we also consume at prodigious rates.

Our hankering for subtropical food provides a handy key to some of our other gustatorial peculiarities. It explains why, for example, we think a drink is somehow incomplete if it doesn't have a wedge of lime in it. Why cooks say nothing brings out the flavour of virtually any dish like a dash of lemon juice. Why we sprinkle pepper over everything we eat. Why we drink so much tea and coffee. Why salsa has replaced ketchup as our bestselling condiment. Why babies' first solid foods are (usually) rice and bananas. Recently, with a house full of youngsters, I put out two pitchers, one of orange juice and another of cranberry juice: the cranberry juice remained untouched. All these seemingly random food choices have something in common; they all involve subtropical plants.

We have, as Darwin might have put it, the habit of sub-tropical foods, a serendipitous habit, perhaps, but a long-standing one that we have chosen not to break. "The degree of adaptation of species to the climates under which they live," Darwin wrote in *The Origin of Species*, echoing poor Klácel's conclusions, "is often overrated." Habit, he said, "is hereditary," and heredity is accidental. A plant or an animal can be moved from its natural climate and can survive and be fertile ("a far severer test," Darwin noted) so long as its new climate is within its habitual range of tolerance. After that, it is only by the process of natural selection – not the handing down of acquired characteristics, but the random success of accidentally mutated genes – that it becomes permanently acclimatized. Darwin also cited the case of the Jerusalem artichoke, "which is never propagated in England by seed," as proof that acclimatization can only be effected by generation upon generation of selecting seeds from only those individual plants that survived the English climate, which no one bothered to do because, I suggest, the Jerusalem artichoke is not a subtropical plant.

Darwin called propagating by selective breeding "artificial selection," which is not random, not accidental, and not natural. It is how we got from Scottish Highland to Canadian cattle, and from Red Malay game hens to Rhode Island Reds. It creates new breeds, but it does not produce different species. Breed a Dalmatian (*Canis familiaris*) with a chihuahua (*Canis familiaris*) and whatever mid-sized, black-and-white yapper you get will still be *Canis familiaris*. Ditto *Felis catus*, *Equus caballus*, and *Homo sapiens*. In Canada, if we wished to adapt to our new environment by artificial selection, we would allow only those individuals who show a marked tolerance for cold climates to breed, or maybe those who prefer Jerusalem artichokes to potatoes, or who think maple syrup is a better sweetener than cane sugar, or those children who *take* the cranberry juice, and think cod-liver oil is a wonderfully refreshing drink. This would have to go on for many, perhaps thousands, of generations, many more than could be observed within the lifetime of one obscure, cantankerous monk living in a remote Moravian monastery, who, incidentally, was relieved from his post as gardener in 1847, in part because of his decidedly non-Augustinian belief that the physical world was a projection of a deeper spiritual reality.

We aren't going to artificially select for a perfectly adapted northern subspecies of *Homo sapiens*, at least not consciously: our physical world will remain a projection of our subconscious wish to be back in the subtropics. As Klácel's work demonstrates, real progress is always accidental. It is most likely that no one would be aware of Klácel's painstaking research at all if, one day in 1843, another monk from the same monastery with a similar interest in gardening hadn't wandered over to Klácel's little patch and asked him what he was up to. That second monk's name was Gregor Mendel.

REQUIEM FOR A RARE BIRD

At the end of a particularly cold February, I came south after spending the better part of four months in the Yukon, which explains why I was thinking about acclimatization, the propensity of a species to tolerate sudden drastic changes in climate. I made a brief stopover in Vancouver, where tulips were up and cherry trees were blossoming, where thoughts of acclimatization often lead to a yen for adaptation: I could get used to this, I thought.

Acclimatization intrigued Darwin as an engine of evolution. The progression seemed to be: acclimatization (not dying in a new environment), adaptation (developing physical traits that optimize survival in the new environment), and evolution (passing those traits along to the majority of individuals in future generations). He had observed that some individuals within a species tolerate sudden changes in climate much better than others. Take a handful of, say, rhododendrons from high in the Himalayas (as botanist Joseph Hooker did in 1850), plant them along the roadsides of England, and some will grow as though nothing had changed while others are suddenly rendered "unfit" (to use the phrase invented by Herbert Spencer, not Darwin, although Darwin approved of it and used it himself in later editions of *The Origin of Species*) and die off. How is it that some plants take climate change in stride while others of the same species, from the same Himalayan slope, pack it in? To Darwin, it was "an obscure question."

There was an instructive example of acclimatization in Vancouver: the city's unique population of a bird known as the Crested mynah. I lingered there a few days to observe it. And also, as it turned out, to observe acclimatization's opposite: extinction.

The Crested mynah (scientific name *Acridotheres cristatellus*, which means crested locust-hunter) is a largish blackbird, about the size of a robin, with small white markings on the underside of its wings (barely noticeable when the bird is not flying), a thick yellow beak, and a tuft of black feathers between the eyes, where the beak meets the head. It is not native to North America; it was introduced to Vancouver from southeast Asia in the late 1890s. There are several murky legends about how this came about, most of them involving Chinese immigrants and birds in fragile bamboo cages. One tells of an exasperated sea captain who, tired of the mynahs' constant chattering, smashed the cages and released the birds as soon as he caught sight of land. The truth is probably more prosaic. Because of their mimetic skills, mynahs – not just the Crested variety but also the more aggressive Indian hill mynah – were popular in Vancouver from the end of the nineteenth century until well into the post-war period. An advertisement placed by the Vista-Variety Store in the Vancouver *Sun* in April 1958 notified customers of a shipment of Indian hill mynahs: "Young, tame birds 3–4 months old, finest talking strain. $45 each f.o.b. Victoria. Limited quantity. Free advice on how to feed and train these wonderful talking birds." Hundreds were imported, some escaped or were released.

However the bird's introduction came about, there were feral Crested mynahs swooping over the city streets by 1897, and in 1904 it was formally identified as a local breeder – citizenship status for birds. Despite the drastically different climate, the population soared. By 1920, there were twenty thousand Crested mynahs in the city. By 1935, its success so alarmed the United States Department of Agriculture that that august body issued a special report urging that "every precaution should be taken to check the spread of this species and prevent its spread into the United States." Crested mynahs were the suspected terrorists of the 1930s.

The USDA needn't have worried, for the mynah never ventured very far from Vancouver. They couldn't make it over the mountains. But even had the birds spread like Japanese knotweed the USDA would by now have relaxed its vigilance, for the Crested mynah has declined steadily since 1950. From its 1920 high, the Vancouver population plummeted to 906 by 1971; in 1980 the figure was 630. The Christmas count in 1985 turned up only 98 mynahs. (In 1983, five Crested mynahs were reported in Dade County, Florida, in the vicinity of the Miami Airport – another last-minute release, perhaps – but by 1989 they were gone.) By 2002, the numbers were even more alarming. Just before my visit, a brief article in the *Globe and Mail* declared that the total number of Crested mynahs in Vancouver had dwindled to "seven, or maybe five." I realized that if I wanted to see a Crested mynah without having to fly to Taiwan, I had better hurry.

The largest single roost of mynahs in Vancouver, comprising more than one thousand birds, had been located at the corner of Cordova and Carrall streets, more or less in the city's downtown core, and that's where I went first, on the chance of seeing a remnant of that enormous flock, or at least to get an idea of the kind of place they once preferred. Cordova and Carrall is an unspiffed corner not far from Gastown. It was noon, still cold in the shadows, a faint warmth when the sun came from behind a cloud. There wasn't much bird life in evidence. The massive buildings had seen better days. Some scaffolding blocked part of the sidewalk. I stood and looked up at the Lonsdale Building, an imposing, three-storey edifice built in 1889. The upper windows were blank, the ground floor occupied by an Army & Navy outlet. A solitary crow landed on one of the upper windowsills and eyed me as though waiting for me to drop a sandwich or a pizza crust. One of the shops on the corner was a takeout chicken joint, and Crested mynahs were once often found scavenging around such establishments – not, as was supposed, because they liked junk food, but because they ate maggots, thus doing the city a valuable service. I poked hopefully around a dumpster behind the chicken place, but there were no birds of any kind on the ground. High above, a few gulls wheeled up from the harbour, and a squadron of pigeons flew over in formation, as though rehearsing for an air show. But that was it.

I walked up a block to Hastings, where a group of street people wrapped in grey blankets sat cross-legged on the pavement, their backs against a formerly grand bank building that had evolved into the Treasure Island Bargain Centre, its treasure of odds and ends spilling out onto the sidewalk like costume jewellery from a battered trunk. More pigeons strutted about outside, pecking at a handful of seeds someone had tossed them. No mynahs. At one point a small, black bird flew down from a leafless tree and joined the pigeons, but it was a starling. I glared at it for a minute, then drifted back to Cordova and continued down to Water Street, the next corner, and the heart of Gastown. Here the storefronts tended to cafés and curio shoppes. Although it was early in the season for tourists, a few sidewalk tables had been set out, so I sat at one across from the statue of "Gassy Jack" Deighton, Vancouver's first hotelier, ordered a decaf Americano, took out my binoculars, and watched the starlings.

The decline of Vancouver's Crested mynah population began with the arrival of starlings in 1950. The European starling (*Sturnus vulgaris*), closely related to the mynah, is another introduced species. Fifty of them were let loose in

New York's Central Park in the 1880s by a drug manufacturer named Eugene Schieffelin, who'd been struck by the dubious notion that North America should have every bird mentioned in the works of Shakespeare. There are now an estimated 50 million starlings in North America, devouring grain and ousting native songbirds from their preferred nesting places. It was the starling experience that turned the USDA against the Crested mynah in 1935.

Ironically, starlings helped ease the USDA's concerns. They ate the same foods and promptly took over all the ideal nesting sites – in Vancouver, that meant the unheated eaves of wooden houses. The weather conspired with the starlings to doom the city's mynahs. To a bird from southeast Asia, anything but a year-round 25 degrees Celsius is cold. Female Crested mynahs usually lay clutches of five to seven eggs, of a colour registered as "light Niagara green," but because they come from relatively tropical climes they don't feel obliged to sit on them for twenty-four hours a day, as more northerly birds – like starlings – do. Mynahs, male and female, spend only about 50 per cent of their time incubating their eggs. The rest of the day they are on the lookout for food. This meant that in the cooler climate of Vancouver on average only two of the eggs in the clutch would actually hatch. It also meant that the nest site was undefended half the time, and after 1950 a lot of mynahs must have flown home from the chip wagon to find a pair of starlings settled in where their own nest used to be.

Vancouver's Gastown turned out to be a good place to think about acclimatization. Up on Hastings, the homeless crouched on the sidewalk or on car seats propped in unnumbered doorways, or cruised the darkened alleys wearing Sorel boots and zipperless winter coats with the hoods up. Here on Water Street, two blocks away, cyclists in yellow Gortex jackets and Spandex shorts basked in the sun, sipping lattes. Both groups seemed comfortably in sync with their worlds. Who were the long-term survivors? Which group was adapting?

After an hour I'd had my fill of Gassy Jack and cold coffee, and took a taxi to an address on the other side of False Creek. Local birders had started a Crested mynah hotline in the early 1990s, and word soon got out that the species' last stronghold, their Alamo, was a small, red-brick building at the corner of First and Wylie, the offices of the Best Janitorial & Building Maintenance Company. As I stepped out of the taxi beside the south entrance I heard some shrill peeping from a row of cedars flanking the building. I spent a few minutes poking among the branches, just in case, before finding a cluster of house sparrows celebrating the springlike sunshine against the brick wall.

When I emerged from the shrubbery, I found a man near the entrance, leaning against the wall, smoking a pipe and reading a book. He looked up as I came out.

"Looking for our mynahs?" he said.

"How could you tell?" I asked, brushing cedar fronds off my shoulders.

He grinned. "They never go in the trees. And they aren't here right now. They usually don't show up until later in the afternoon. They were here yesterday."

His name was Cliff and his office was on the third floor, second from the right. His window was at the same height as two exterior light fixtures, one sticking out of the brick just to the left of his window and the other two windows over, above the cedar hedge. Between Cliff's window and the far light fixture was a hole in the brick wall where a third, central light fixture had once been. Cliff pointed out these features carefully. The mynahs perched on the two extant light fixtures, he said – I could see a fan of white guano on the wall under each metal pole – and they nested in the middle cavity where the third fixture used to be. Cliff could watch all three sites from his window, and was thus probably the most experienced observer of Crested mynah behaviour in North America.

"It gets very hot inside during the summer," he said. "Even in early spring this south wall gets a lot of sunlight. That's probably why they like this building."

I asked him how many mynahs he'd seen this year. "We're down to three," he said sadly. "Last year we had five: two males, a female, and two hatchlings. The females don't sing as much, and their crests are smaller. I take it the two hatchlings were females, and something bad happened to the males over the winter, because now there are only three, and all three are female."

When Cliff returned to work, I crossed Wylie Street and fixed my binoculars on the nest cavity. The entrance seemed to slant up, probably leading to a larger hole deeper in, where warmth from the building would nurture the eggs. This would give these birds a slight advantage over other mynahs. Mynahs carry dried grass and even paper and bits of plastic into their nests for warmth and comfort; Crested mynah nests also, according to a 1950 report, "invariably contain a snakeskin."

Here's how Darwin thought acclimatization led to evolution: when the mynahs started reproducing in Vancouver in 1904, most of them spent 50 per cent of their time incubating the eggs, but a few would spend 53 or maybe 55 per cent. Those that spent more time on their eggs would have a slightly better chance of producing young. Perhaps the 50-per-centers would successfully raise one chick, while the 55-per-centers would raise two or three. Gradually, the 50-per-centers would disappear, and the population would be made up entirely of 55-per-centers. Among this population, a few 60-per-centers might emerge, and gradually they would ease out the 55-per-centers. Over time, the Vancouver Crested mynahs would have acclimatized so well they could be considered to have adapted; they would constitute a new subspecies that sat on their nests much longer than the original Asian population. Before that happened, however, the starlings arrived.

I hung around, but no mynahs appeared. When the sun dropped behind the Maynard's Auctioneer building behind me, it began to get really cold. At four o'clock the police officers in the Canine Division compound across First changed shifts. A few minutes later a thin man in a light jacket pushed a shopping cart up to the Best Janitorial dumpster and started rummaging for plastic bottles and aluminum cans. He whistled as he worked. He made me think of mynahs, chirping and foraging in the city's trash cans for a living. A lot of us can live off what others throw out, as long as the competition keeps away. This man told me a friend of his was going to give him a brand-new, twenty-eight-inch TV set – all he had to do was go and pick it up, which he planned to do the next day. "You don't have a car, do you?" he asked me. I said that regrettably I did not. I was just passing through. When he left, I looked up at the nest cavity and the light fixtures one last time. Is this what extinction looked like, I wondered? All day watching an empty hole in a brick wall?

The next day I arrived shortly after noon and stayed for an hour, this time with my friend Ed Good, who is an excellent birder and a fine conversationalist. We sat in his car with the heater on, and he told me about walking around Vancouver only a few years ago and seeing Crested mynahs on the sidewalks, mixed in with starlings, which to a casual observer they resemble. "You really had to look closely," he said, "but in a flock of fifty or so you usually saw two or three mynahs." It was a tactic; the fraternizers may have been the last mynahs to survive.

After lunch, Ed went back to work, and I returned to the Best Janitorial building and settled in for another vigil. This time, I was rewarded. At 4:20 two black birds flew toward the building from the direction of the Cambie Bridge and landed on the light fixtures. At first I thought they were starlings – their white wing flashes were not very visible when their wings were folded – but when I looked through my binoculars, I could see the tuft of feathers at the point where the tops of their beaks met their foreheads. They had large beaks, and their eyes were bright red. They looked about them with nervous inquisitiveness. Two female Crested mynahs. I could barely contain my excitement. I waved at Cliff's window, between the two light fixtures that held two-thirds of the entire North American population of Crested mynahs.

Darwin thought that acclimatization would gradually lead to adaptation: as Vancouver's Crested mynahs, say, became more habituated to a colder climate they would adapt to it physiologically as well, and over vast stretches of time they would evolve into a new species. "Let this process go on for millions of years," he wrote, "on millions of individuals . . ." But researchers now think that adaptive changes could take place more quickly. Jonathan Weiner, for example, in *The Beak of the Finch*, suggests that adaptation "need not be as gradual as Darwin imagined." Citing Darwin's finches on the Galapagos Islands, whose beaks adapt to differences in seed availability from one year to the next, adaptive changes could take place within a few generations. Still, adaptation has to persist for a long time before it becomes evolution. It's as though the species, having survived one climate shift, waits around for a while to see if the climate is going to shift again before committing itself to an evolutionary plunge.

It's possible that, left alone for a few more years, Crested mynahs would have become better acclimatized to Vancouver's climate, might even have figured out a way to spread beyond the mountains and thus fulfill the USDA's worst fears. Sitting on a nest for a crucial extra hour or two a day is hardly a huge leap forward. A few years ago, two University of British Columbia evolutionists, Craig Benkman and Anna Lindholm, conducted an experiment on crossbills, a type of seed-eating finch whose bills curve sharply at the tips and do not meet, like a pair of badly aligned nail scissors. The birds have adapted to opening a particular kind of hemlock cone; when Benkman and Lindholm clipped their beaks, the birds could forage open cones, but were unable to open tightly closed cones. As the beaks grew and became more crossed, the birds were able to open closed cones again. This suggested that beak alterations had occurred gradually in nature, and that the birds would have had to adapt in many subtle ways to take advantage of the changes. Citing the UBC experiments, Weiner posits that crossbills with slightly altered beaks would have needed to refine their instincts for cone hunting, learn to recognize new types of food, develop new muscles to operate their new beaks, and so on. These physical changes would eventually lead to social and reproductive changes (as females chose mates with well-adapted beaks), and before long the world would have a whole new species of crossbill. Given more time, something similar might have happened with Vancouver's Crested mynahs.

But it didn't. Starlings happened instead. As I watched, the two Crested mynahs left their perches on the light fixtures and flew to a nearby telephone cross-tree, obviously hoping to roost for the night. Within five minutes they were assailed by three starlings. One of the mynahs scooted along the beam, chasing two of the starlings off, but the starlings merely flew up onto a wire and then returned. Before long two more starlings arrived, and both mynahs moved grudgingly back to their light fixtures, perhaps to protect their nest, but it seemed more as though they had just given up, realized that there was nowhere else for them to go, that wherever they tried to roost they would be ousted by starlings. Not violently, not aggressively, just edged out by sheer force of numbers, made uncomfortable, unwanted, forced into retreat.

A year later I asked my friend Ed about the mynahs. He told me the last one had died. He heard about it through the hotline. Someone had found the bird on a sidewalk, possibly on Wylie Street, outside the offices of the Best Janitorial & Building Maintenance Company. Whoever it was collected the body and buried it somewhere in Stanley Park. "It isn't extinction," Ed consoled me. "It's extirpation. There are still a lot of Crested mynahs in Asia."

Maybe so, but it feels like extinction, like an end. And it feels like failure – failure to adapt, failure to evolve. It is too close to our own situation to be comfortable, for we, too, have failed to adapt or evolve. We may be acclimatizing, after a fashion, but the hordes of starlings are coming down from the mountains.

On Walking

What is it, what is it,
But a direction out there,
And the bare possibility
Of going somewhere?

— THOREAU, "The Old Marlborough Road"

Late in the morning of April 27, 2002, Charles Wilkins set out from his home in Thunder Bay, Ontario, with the idea of walking to New York City. It wasn't a last-minute vanishing act; he did make some preparations. He got a cellphone. He bought some gear. He convinced a friend, George Morrissette, who is a poet, to "more or less keep pace" with him in a van. He threw a few necessities into the van — a laptop, some warm clothes, extra shoes, tins of sardines — and set off, with Morrissette driving on ahead to wait for him somewhere far, far down the road. He soon realized that there were a few things he hadn't done enough of. Training, for example.

"On my first day out," he told me while still en route, "I walked twenty-five kilometres and was wobbly by the time I stopped. The next day my knee hurt so badly I had to finish my walking going backwards!" But the body, he said, "adjusts marvellously." It took only three weeks of walking for him to become "totally comfortable," during which time he upped his daily progress to thirty-five kilometres, dropped his weight to 150 pounds, gained two inches on his calves, hardened his buttocks, and shed a dozen layers of skin from his soles. "My feet," he reported, "have adjusted to the point where, when a layer of skin is ripped off, there's another waiting serviceably underneath, ready to go."

His motivation for taking the big walk, apart from "simple curiosity about this most ancient and fundamental means of transportation" – more than any other attribute, it is upright walking that distinguishes us from most other vertebrates: even birds, though bipedal, walk with their spines parallel to the ground – was to prove to the rest of us that we have become a sedentary species, or rather a species that, thanks to technology, moves incredibly fast while remaining physically and mentally in one place. "While jet travel and the Internet shrink the planet," he said, "walking expands it." By walking, he slowed the world down to a pace "at which noticing becomes not just possible but unavoidable."

Such a pace is too slow for most of us. As a rule, North Americans don't walk. I mean really walk, as Charles walked, or as they walked, say, in Chaucer's England, when April with its sweet breath reportedly made everyone long to go on pilgrimages and talk one another's ears off. Even eighty years ago walking was more common than it is now. Walking tours were a commonplace, like today's bus tours. In 1927, the novelist V.S. Pritchett set off, on a whim and on his own, to walk across Spain. He wanted, as he recorded in his first book, *Marching Spain*, to know "the monotony of that burned-up country, the dumbness of its cottages and taverns. It would be unpleasant for my body," he shrugged, "but for the soul it would be ennobling."

Nowadays, as Wilkins noted, the car drive and the hard drive have yanked the legs out from under us. He might have added that, while motorized conveyances have brought distant places closer, they have also robbed us of the nearer destinations: we think nothing of driving an hour to see a movie but wouldn't dream of walking for an hour to see a friend. A study conducted recently at the University of California determined that the average American walks less than seventy-five miles a year, or about one thousand steps a day. As Bill Bryson, author of *A Walk in the Woods*, notes, "I rack up more mileage than that just looking for the channel changer."

Bryson's plan to test the level of wimpiness to which we have descended was to spend two summers racking up the better part of two thousand miles walking the Appalachian Trail from Virginia to Maine. About two thousand hikers attempt it each year, of whom fewer than two hundred make it all the way. Not to detract from their achievement, but completing the Appalachian Trail, or the Yellowstone-to-Yukon Trail, or the still-on-the-drawingboard Trans Canada Trail, is not walking but hiking, and hiking differs markedly from what Charles Wilkins did, which was closer to answering American naturalist John Muir's call to "throw a loaf of bread and a pound of tea in an old sack and jump over the back fence." A hike is more like a business trip than a vacation. It has that purposeful, heads-down, chalk-up-the-kilometres feel to it that walking lacks. Chaucer's pilgrims, though they walked with a purpose and many of them carried sticks, cannot be said to have been hiking. They told each other stories as they went. In fact, they told so many stories they never made it to Canterbury.

Thoreau would have said they were sauntering. In his essay "Walking," he traced the origin of the word *saunter* to pilgrims making their way to the Holy Land, the Sainte-Terre, and so were called "Sainte-terrers," which became saunterers. That is, of course, a wild conjecture – the OED admits the word's origin is "obscure" (an OED euphemism for anybody's guess, without acknowledging Thoreau's) – but it does catch the sense of a pilgrimage as something done more for the sake of walking than of getting anywhere.

Thoreau believed that walking freed the soul, allowed the mind to range as it could not when confined to a cabin, a cab, or a coach. "I think that I cannot preserve my health and spirits," he wrote, "unless I spend four hours a day at least – and it is commonly more than that – sauntering through the woods and over the hills and fields, absolutely free from all worldly engagements." His essay is a record of the many and varied and often unrelated thoughts that flitted through his mind as he walked, composing his own version of the

Canterbury Tales. He never walked on roads, he said, and always instinctively headed west when he left his cabin, for "the West of which I speak is but another name for the Wild, and . . . in Wilderness is the preservation of the World." Humans require daily sustenance from nature: "Every tree sends its fibres forth in search of the Wild," and we should do likewise.

Although Bryson was hiking, he agreed with Thoreau. The restorative power of walking, he wrote, comes from the fact that, while doing it "you have no engagements, commitments, obligations, or duties; no special ambitions and only the smallest, least complicated of wants; you exist in a tranquil tedium, serenely beyond the reach of exasperation."

When Thoreau proclaimed himself a walker, he meant something a little more demanding than going for an after-dinner stroll. "If you are ready to leave father and mother," he wrote, "and brother and sister, and wife and child and friends, and never see them again, if you have paid your debts and made your will, and settled all your affairs, and are a free man, then you are ready for a walk." Was that how he prepared for the trip he described in *A Yankee in Canada*, published in 1866, for which he walked from Montreal to Quebec City, a distance of some three hundred kilometres, in a week? One suspects not. He made the journey in much the same way that walking-tourists today amble about the Italian or English countryside, sticking to the roads, light day-bags over their shoulders, and eyes sharply peeled for five-star inns. When a coach offered itself, Thoreau happily hopped aboard.

Getting up on our hind legs was, biologically speaking, an interesting posture to adopt. Evolutionists argue about what prompted it. There are as many theories as there are theorists. Even a brief rundown of them is illuminating (my objections to them are in brackets): walking upright made it easier to carry food and babies (all animals carry food and babies); it evolved from wading and swimming (then why didn't other aquatic attributes, such as webbed feet?); it allowed us to follow migratory herds across the savannahs (you mean, the way lions do?); it was more energy efficient than walking on four legs (then why don't more animals do it?); it exposed our bodies to less sun, thereby allowing us to colonize hotter habitats (so would smaller heads and keeping our bodily hair, and some of us can't even keep the hair on our heads); it exposed the male genitalia, which was either attractive to females (!) or frightening to rival males (!!).

The one I like best, though it has been unpopular for two decades, holds that, as early hominids emerged from the forest onto the savannahs, they needed to stand upright in order to see above the tall grass. One may wonder why other grassland species — lions and hyenas and gazelles and such — remained staunchly quadrupedal, but perhaps having started out as tree-dwellers, we had become accustomed to seeing greater distances, to being able to situate ourselves with reference to a horizon, or a distant landmark, and that is what we missed when we sauntered out onto the prairie. Little knolls and the occasional shrub didn't do it for us. We needed perspective.

However we acquired bipedality, the benefits soon multiplied. Most evolutionists credit our large brains to the complex neurological demands imposed upon it by upright walking. Seeing farther meant seeing more; the brain had to deal with a vast increase in peripheral input, had constantly to assess the possible meaning of vaguely perceived movements and shapes. Our binocular vision sharpened, turning us into predators – usually a bigger-brained occupation than cud-chewing. (Predatory species, with the exception of some fish, are binocular; that is, with both eyes facing forward and thus providing depth perception. Prey species tend to have eyes on the sides of their heads, to see more of what's happening all around. Think of eagles and chickens, or wolves and sheep.) Standing upright also meant hearing more, which required extensive renovations to our auditory chambers.

Many dinosaurs were bipedal, particularly those that evolved into birds, without developing Mensa-sized brains. Why not? One possible answer is that they didn't have much in the way of front legs before they started walking on their hind legs. Most of the big bipedal carnivores, the *T. rexes* and their raptorial kin, had spindly little front legs that were so short they couldn't reach their gigantic mouths. It's hard to think what use they were. Too short for feeding, no good for balance, useless for holding down an opposite number during battles or sex. Such forelimbs hardly required a dinosaur to do much more thinking than it did when it was horizontal. But the forelimbs of primates were at least as long as their hind-limbs, and standing upright left those long front legs and hands free to develop into tool-holding appendages, which allowed the brain to devise splendid new uses for such things as sticks and rocks (with which materials we still make most of our houses). The palms of our hands became almost as sensitive as our tongues. Our brains swelled to accommodate all this new input from the five senses.

There were, however, trade-offs. Walking on two legs is riskier than walking on all four, just as riding on two wheels is wobblier than riding on four. John Napier, an evolutionist who argues for the early development of bipedality in humankind (before larger brains), writes that "human walking is a unique activity during which the body, step by step, teeters on the edge of catastrophe." It may not be such a bad thing that we teeter on that edge for only 315 metres a day; couch potatoes may be the saviours of our species – unless, of course, we need an occasional dose of catastrophe in our lives, to keep the adrenalin flowing. There is much in history to suggest that that is the case.

If so, walking would do it. It's a nail-biting, bone-jarring activity. After 3.7 million years of near-disaster, we have arrived at a better bone structure for bipedalism, but we still haven't got it quite right, and there have been attendant disadvantages. We have widened our hip joints and brought our knees together (chimpanzees have narrow hips and bowed legs, which accounts for their awkward gait on land) and the pelvis, as Rebecca Solnit observes nicely in *Wanderlust: A History of Walking*, "has tilted up to cradle the viscera and support the weight of the upright body, becoming a shallow vase from which the stem of the waist rises."

But, as physiotherapists know, wide hips and narrowed knees don't seem to work well together. The patellae, or kneecaps, wander around, slip sideways, tilt upward, and tendons and ligaments have a half-life of about twenty years, which is why knees are often the first things to go in athletes such as hockey players and skiers. By the age of fifty, most of us begin to stiffen up in the knees, and by the time we reach our allotted threescore years and ten, shoe inserts, knee and hip replacements, and patella manipulations are our common lot.

But the real adrenalin flows during childbirth. The brain of a human infant at birth is already as large as that of a full-grown chimpanzee, and yet the birth canal in humans has become smaller over evolutionary time, a consequence of our upright stance, making childbirth – the thing animals are supposed to be best at – a difficult, painful, and sometimes fatal process for both mother and child. We gestate our children longer than other primates, because our big brains take longer to develop; but in doing so we jeopardize our chances of surviving birth. This seems a contradiction. Our large brains are both our liberation and our limitation.

We seem to thrive on contradiction. Charles Wilkins observes that walking gave us larger heads, which made us more intelligent, which allowed us to invent ways to avoid walking. Thoreau made his living as a surveyor, pacing out and parcelling up the land that he would have remain free and unencumbered. In "Walking," for all its Whitmanesque enthusiasm for wilderness – "I believe in the forest, and in the meadow, and in the night in which the corn grows" – he also suggested it was by taming the wilderness that humankind achieved its ultimate purpose: "I think that the farmer displaces the Indian even because he redeems the meadow, and so makes himself stronger and in some respects more natural." We don't normally think of Thoreau as a spin doctor for land development.

All of which goes to show that, while walking, the mind wanders with the feet, and both can be errant if they are so inclined. Many ancient and modern philosophers walked in order to set their thoughts to roaming. Greece had its Peripatetic school of wandering scholars, whose loyalties were more to ideas than to city-states — "an attachment," Solnit notes, "that requires detachment." Thomas Hobbes carried a walking stick with a built-in inkwell, in case he had a brainstorm during his afternoon stroll. Einstein claimed to have done his best thinking while walking on a beach. Wordsworth and his fellow Romantic poets regularly stalked the Lake District, though they took the train to get there (Bill Bryson remarked sourly on his neighbours in New Hampshire who drove the few blocks from home to their exercise classes, and I have a friend, a carpenter, who recently worked on a house in which the occupants took an elevator up to the fourth floor, where they kept their StairMaster).

"I nauseate walking," William Congreve, the eighteenth-century playwright, has one of his characters opine. "'Tis a country diversion; I loathe the country." I live in the country, but I do not find walking a country diversion. Country people rarely walk, and those who do are noticed. A neighbour passes our house regularly and often stops to chat (and, I suspect, catch his breath); he has been ordered to walk three kilometres a day by his doctor. It isn't a walk, then, it's a forced march. "But I enjoy it!" he says, surprised, as I imagine psychiatric patients might have said about cold baths.

I do most of my walking in the city, especially in unfamiliar cities. A few years ago, finding myself with four days' unexpected leisure in Buenos Aires, which is a very large city, I walked each day from early morning until after dark, crossing its extraordinarily wide *avenidas* and exploring its hidden *ruels* and passageways, making some of its secrets my own. I have done the same in dozens of cities: Beijing, London, Florence, Oslo, Helsinki. My wife, Merilyn, and I wore ourselves out walking in Paris, doing what Americans call "rubber-necking," or taking in the sights. The philosopher of city walking, Walter Benjamin (who called Paris "the capital of the 19th century"), employed the term *flâneur* to impart the same sense of alert yet aimless wandering that Thoreau meant by "saunterer," and that's pretty much what I am in a strange city, a flannerer (to anglicize an old Norman form of the same verb). Baudelaire, in three poems dedicated to the exiled Victor Hugo ("The old Paris is gone / the form of a city / changes more quickly, alas, than a mortal's heart"), was writing as a flannerer. Paris was made for flannery, but any large city will do. Stephen Daedalus, the hero of James Joyce's *Ulysses*, was a Dublin flannerer – "Am I walking into eternity along Sandymount strand?" Apparently, one can flanner in a city one knows well. Why not? Thoreau could saunter in the same woods every day and always have a different string of thoughts, and I doubt that Einstein needed a new beach for each new equation. I have often wondered at my disinclination to saunter in the country, given my propensity to flanner in the city. Driving to the city in order to go walking seems like a contradiction to me, like taking an elevator up to the StairMaster.

Contradictions, however, are the fertile ground from which decisions arise, and as V.S. Pritchett remarked, "All the great decisions in life are made by the legs." A return to walking may be the greatest decision we can make, the best way most of us have of slowing down our lives and falling into step with the natural rhythms of nature. Charles Wilkins hoped to get to New York in September or October, before it became too cold, but he wasn't worried about it particularly. New York wasn't what he was looking for. "One thing I have noticed," he said as he strolled past Sault Ste. Marie, "is the way in which time seems almost to have been created for me, to have been returned to me, as I walk."

That's a better discovery than New York.

NOCTURNE

Hanging in the National
Gallery in Ottawa is a painting by William Kurelek that
looks, from a distance, like a large square canvas painted
black. Upon closer inspection, it is a night scene. In the centre
is a large, ploughed, dark field, and to the right, a water-filled
ditch running down the side, with a footbridge crossing it.
You can just discern a group of people on the bridge, rushing
across, seemingly panic-stricken (one man has lost his footing
and is falling into the ditch). It seems to be a religious gath-
ering, all are carrying copies of *The Watchtower*, and they are
looking over their shoulders at something that is happening
in the field, something terrifying. When you follow their gaze
you see it is an owl, white, spectral, its eyes and claws wide

open, plunging down through the darkness toward a family of tiny, white mice. The mice, too, have seen the owl, and are scampering off to the left, as panicky as the people on the bridge. The painting is entitled, "Blind Leading the Blind."

Carl Jung believed that when an artist is fully engaged in his or her work, he or she is tapping into some kind of collective subconscious, and when a viewer sees and responds to that work, he or she is drawing from that same subconscious source. So, what is going on in this painting? It is night. There are two groups, the humans and the mice. The mice are frightened, but so are the humans. The mice have good reason to be frightened – they are about to be dive-bombed by an owl (although in life mice probably never see the owl that gets them). But why are the people also terrified? They are in no danger, except that created by their own panic. And if both groups are blind, who is leading whom?

It must be the night itself that has frightened the people, a night filled with unseen, or unseeable, terrors. Hence, probably, the religious motif. Night and religion are commonly linked in Western art. Christopher Dewdney, in his book *Acquainted With the Night*, notes that one of the earliest paintings of a night scene is an Egyptian image of the sky goddess Nut arched over the earth, who is both her brother and lover. Night is the time of spectral beings. It is no accident that Hamlet's fateful encounter with his father's ghost takes place at midnight. "I am thy father's spirit," says the Ghost, "doomed for a certain term to walk the night." There is no logical reason why ghosts should be more active after sundown than before it, but Shakespeare, who knew a nighthawk from a handsaw when he saw one ("handsaw," by the way, is a corruption of "herronsew," an old name for a juvenile heron), also knew how to evoke the classic involuntary fright responses from his spectators: goose flesh, widening of the eyes, stiffening of the neck hairs (or, as Shakespeare himself more eloquently put it, scenes that would "freeze thy young blood, Make thy two eyes, like stars, start from their spheres, . . . And each particular hair to stand on end, Like quills upon the fretful porpentine"). Setting the scene at night, as directors of modern horror films instinctively know, is guaranteed to make the audience squirm in their seats. "'Tis now the very witching time of night," Hamlet says, "When Churchyards yawn and Hell itself breathes out Contagion to this world."

Witches, like ghosts, could (and did) just as easily practise their black arts during the day, so why do we have them cavorting by candlelight? It may be because early sorceresses achieved their ecstatic visions by dosing themselves with the juice of the night-blooming plant *Datura stramonium*, a fragrant, hallucinogenic member of the nightshade family, known to Thoreau as thorn-apple or Devil's trumpet, from the shape of its flowers. Datura contains a number of mind-altering compounds, including hyoscyamine, atropine, and scopolamine, "a tropane alkaloid," writes Wade Davis in *One River*. "If you take a big whack, it brings on a wild, crazed state, total disorientation, delirium, foaming at the mouth, a wicked thirst, terrifying visions that fuse into a dreamless sleep, followed by complete amnesia." Did the Pilgrim Fathers check Salem's gardens for datura? The plant was found growing wild in Virginia in 1676 by a group of British soldiers on their way to put down the Bacon Rebellion; they boiled and ate it as a salad (the British boil everything, even their salads). "The effect," according to the eighteenth-century historian Robert Beverley, who must have disliked British soldiers, "was a very pleasant comedy."

But the real reason we fill the night with terrors may have more to do with our natural history than with religion. The question is: what put those terrifying images in our subconscious in the first place? Kurelek's haunting painting provides, I think, a tantalizing clue. The only blind beings at night are us. Mice can see in the dark as well as owls can hear. We can do neither very effectively. We haven't always been that way; far back in the evolutionary history of our species, we were nocturnal creatures. When we emerged from the night, we traded in night vision for colour vision, and darkness has frightened us ever since.

As mammals, we are descended from an extinct group of reptiles, the therapsids, that crawled the Earth some 400 million years ago. They were the planet's dominant vertebrate life form until dinosaurs took over about 250 million years ago. Triassic dinosaurs were large and cold-blooded and therefore diurnal – they couldn't maintain their body temperature if they moved around much during the night – so as the dinosaurs seized the day, some of the smaller therapsids survived by becoming nocturnal: in paleoeconomic parlance, the night was empty, and the therapsids colonized it. To do so, they had to develop higher body temperatures, which they eventually maintained by becoming warm-blooded and developing hair, and they augmented their reduced eyesight with acute senses of smell and hearing. Sometime in the late Triassic or early Jurassic, some of them became mammals.

Most present-day mammals − like the mice in Kurelek's painting − are still nocturnal, while most reptiles, as well as birds (except for Kurelek's owl), remain diurnal. The big exception among mammals is us, the primates. Prosimians broke off from the main mammalian line about 60 million years ago, shortly after that famous meteorite smacked down in Yucatán and rid the planet of dinosaurs. For some reason − perhaps because the day suddenly became empty, whereas the night had become full of a wide range of competing (or frightening) predators − the primates returned to daytime activity; they became diurnal, and the night became a distant, disturbing memory, a bad dream. One of the ways they were able to do this was by developing a new kind of eye.

The evolution of the eye is a logician's nightmare, full of twists and unknown turns and even, it seems, at least one impossibility. Darwin worried that anti-evolutionists would seize upon the eye to challenge his theory of natural selection: how, he imagined his detractors demanding to know, could such a miraculous organ as the eye have simply developed, stage by stage, from nothing? Surely it had to have been created by God. "If it could be demonstrated that any complex organ existed which could not possibly have been formed by numerous, successive, slight modifications," he wrote in *The Origin of Species*, "my theory would absolutely break down." Anything that apparently leapt into existence without a slew of previous forms preserved in the fossil record could be seized upon to discredit his theory of how evolution works.

His problem, of course, was the complete absence of fossil evidence for any organ's gradual development. Bones and teeth fossilize, and skin and feathers have left impressions in volcanic dust, and so a succession of subtle changes in them can be traced. But soft organs almost never become fossils. There are no dinosaur hearts or lungs to study, alas. (Certainly in Darwin's day nothing of the kind existed; in the past few decades dinosaurs with fossilized skin have been found, and I have seen a Carnatosaur specimen in Argentina with skin-covered muscles, but no eyeballs.) Darwin had to posit the evolution of the eyeball, which began as a cluster of primitive, photosensitive nerves "surrounded by pigment-cells and covered by translucent skin, but without any lens or other refractive body," serving merely to distinguish light from darkness. Step by infinitesimal step, over eons of time, the cluster evolved "into a structure even as perfect as an eagle's eye."

The resultant eyeball is little more than a hollow sack (the sclera) surrounding a photosensitive liner (the retina). The retina is essentially an extension of ganglia from the brain attached to a layer of cells of two types: rods and cones. Photons of light enter the hollow eyeball through a lens fitted into the cornea, are absorbed by the rods and cones arranged along the back wall, and are changed into electrical impulses that pass along the ganglia to the brain, which reinterprets the impulses back into an image.

That sounds simple enough. Digital cameras work much the same way when hooked up to a computer. But this is where it starts to get unpredictable (i.e., human). The rods deal with black and white; the cones with colour. Because the eye is thought to have evolved from cells whose sole function was to distinguish dark from light, thus triggering certain hormonal secretions, you'd think the rods would have come first, with the cones coming along later as the eye evolved to perform more complex tasks. But not so: the cones came first. Those sluggish therapsids could discern colours, maybe more of them than we can. They lost much of that ability when they became nocturnal: they didn't need colour vision any more, because they couldn't see colour at night anyway. That's when the number of cones dwindled and were outnumbered by rods.

Most modern mammals still have more rods than cones; they can see black and white and all the shades of grey, but the only colours they perceive are those on the low-wavelength end of the spectrum: blue and green. They are dichromatic. They cannot distinguish red from green. Despite all the hoopla, matadors might just as well be waving a lime-green pillowcase in front of a bull as a fiery red cape. It's all the same to the bull: it's the indignity of it that infuriates him, not the colour red. Similarly, to a cat, a canary could be white or magenta or puce; cats can only distinguish green from blue. They aren't biologically inclined to hunt birds anyway, given that birds are diurnal and cats nocturnal. Cats are meant to hunt rodents and other small mammals, not birds.

Reptiles and bony fish have retained their original cone dominance. Birds, the reptilian descendants of dinosaurs, are tetrachromats: they can see four colour dimensions, making them able to discern colour ranges from black to white, blue to yellow, green to red, and violet to ultraviolet. They can see colours we would never imagine. What to our impoverished eyes appear as drab, dun-coloured sparrows, which birders disparagingly refer to as LBJs (little brown jobs), may very well appear to other birds as brilliantly bejewelled as a peacock. In fact, we may be seeing only some of the colours of a peacock. Researchers suspect, for example, that species that seem to us to be monomorphic – both sexes appearing identically coloured, like blue jays or crows – are in fact sharply differentiated in the UV spectrum. It is also entirely likely that in birds' ability to see UV light lies the secret of annual migration.

Primates are the anomaly. When prosimians broke off from the main mammalian line by becoming diurnal, they evolved an extra set of colour-detecting cones in their retinae. Re-evolving a character that has evolved out is supposed to be an evolutionary impossibility. Dollo's Law on the irreversibility of evolution (Louis Dollo was a Belgian paleontologist who was the first to realize that some of the bigger dinosaurs were bipedal, thus providing an early link between dinosaurs and birds) has it that once a feature, like gills or quadrupedality, has evolved out, it's gone forever. According to Dollo, flightless birds like the ostrich and the rhea (assuming that they were once flighted and have since lost that ability) will never fly again.

But somehow, Dollo's Law notwithstanding, when the prosimians split away from dichromatic mammals, they re-evolved cone dominance, thus becoming trichromatic primates, which is why we can see high-wavelength reds and yellows (a range that allows us to see 2.3 million different colours). It is unDarwinian to ponder the reasons for this evolutionary about-face – evolution doesn't have a reason, it just happens and species take advantage of it. A species doesn't discover a need for trichromacy, and then set about accomplishing it. Rather, a few members of a species accidentally, through environmental change or chance gene mutation, find themselves with a slight ability to see something more than the basic black and green. Something in the species' habitat favours that ability, and those individuals live longer than their dichromatic siblings. Over many generations, more trichromatic individuals survive than dichromatic ones, and a new species is born. Today, most Old World monkeys, apes, and humans are trichromatic – there are some exceptions, such as bush babies and lemurs – although most New World monkeys, such as owl or night monkeys, are dichromats.

What was trichromacy's advantage? Why was being able to discern red and yellow from green and blue a giant step for humankind? Not surprisingly, several hypotheses have been put forward. One inevitably has it that trichromacy is a sexual aid. In some species of Old World monkeys – guenons, for example – the male's blue scrotum is surrounded by bright yellow fur, a come-hither fashion statement that wouldn't get a rise from a dichromat. Likewise, the females of some baboon species have bright red calluses on their rumps, which are irresistible to trichromatic males. Similarly, male mandrills have blue, violet, and scarlet genitals. None of these sexual signals would be picked up by dichromats; trichromats may use them to distinguish other members of their own species, and to determine when potential same-species mates are sexually mature.

More recent theories focus more on the stomach than on sex: trichromacy, it is now thought, helped diurnal primates discern red and yellow fruit in green foliage. Dichromats have great difficulty distinguishing between unripe (green) and ripe (red or yellow) fruit against a background of trembling leaves. Trichromacy thus made food foraging faster and easier. An Old World primate could spot a tree containing ripe fruit from a greater distance. In *The Island of the Colorblind*, Oliver Sacks visits the South Pacific island of Pingelap, on which almost everyone is born dichromatic. Whether a coincidence or not, Pingelaps were not much disadvantaged by their colour-blindness; Sachs found "it striking how green everything was in Pingelap, not only the foliage of trees, but their fruits as well – breadfruit and pandanus are both green, as were many varieties of bananas." This may explain why most New World monkeys are dichromats – they didn't need trichromatic vision to find food. Sachs quotes a study conducted by J.D. Mollon, a researcher on the mechanisms of colour vision, who noted that Old World monkeys "are particularly attracted to orange or yellow fruit (as opposed to birds, which go predominantly for red or purple fruit)."

These observations inspired a recent study conducted in Uganda, which showed that trichromacy is also useful to leaf-eaters. In Africa, the leaves of up to 50 per cent of edible trees are red before they turn green, a phenomenon known as delayed greening. Young red leaves are easier to digest and richer in protein than the more fibrous mature green leaves, and are also available at times of the year when mature fruit is not. Animals that could discern both young leaves and mature fruit could thus find food for longer periods, and therefore enjoyed a decided advantage over those that could see only green fruit and green leaves.

It is our eyes that made us successful day creatures. But they also robbed us of the night. When we emerged from nocturnality to become diurnal, is it possible that we retained, embedded somewhere deep in our subconscious, the memory of a darkness filled with predators, a night realm we could no longer see, but which we knew to be rife with danger? An ache that religion salved? Freud believed that we invest our nighttime dreams with images gathered during the daytime: "day residues," he called them. Could our new, daytime imaginings be fuelled by a species memory of our nocturnal selves, filled with "night residues"? If so, it might explain why so many of the stories we scare ourselves with take place at night, and contain creatures we cannot ordinarily see. Witches and ghosts, ogres, transgenetic beasts. Such creatures are our night residues.

One of my favourite poems is John Donne's "Nocturnal Upon St. Lucy's Day." Saint Lucy's Day, by the old Elizabethan reckoning, was December 21 (it is now the thirteenth), the winter solstice, the shortest day of the year and therefore the longest night. In his poem, Donne calls it "the year's midnight." It is a singularly dark and melancholic poem. When Donne decides to be gloomy, he can be downright suicidal:

"The world's whole sap is sunk," he writes. "For I am every dead thing . . . the grave / Of all, that's nothing." He never tells us why he is so despondent, although he hints at the death of love. Has his lover died? Has she left him? Or is it all love that is dead, impossible, always tantalizingly out of reach? Whatever is afflicting him, it is definitely something more than an early case of Seasonal Affective Disorder. Like the distant sun, he is "re-begot / Of absence, darkness, death — things which are not."

Or things which are no longer, perhaps. In fixing his despair on that particular day, he hit upon the ideal symbol, one that suggests his poetic imagination went fishing in the Jungian pool and pulled up something from our evolutionary past. St. Lucy is the patron saint of those afflicted in the eyes, especially of the blind. She lived in Syracuse, Sicily, in the fourth century. In one version of her story, when a nobleman fell in love with her because, he claimed, of the beauty of her eyes, she plucked them out, Oedipus-like, and gave them to him on a platter. "Now let me live to God," she said, and he apparently did. In Christian art she is represented with a palm leaf in one hand and in the other a platter, on which roll two sightless orbs. Like loveless Donne, and like all of us at one time in our evolutionary history, she dwelled in perpetual night.

OF BEANS AND BEARS

No one can say that Henry David Thoreau did not love nature. His was an unabashed, euphoric passion: "I love the wild not less than the good." But there are, to me, certain oddities in the way he expressed his feelings for the wild, even to himself. He liked to hunt and fish, for instance: "The wildness and adventure that are in fishing still recommended it to me." In hunting he preferred the technique known as "still hunting" – tracking and waiting and acquiring a thoroughly intimate knowledge of the ways of the animal being hunted, so that hunting became a kind of field exercise in natural history – but he was not above relishing the blood-sport aspect of it, the thrill of the kill. There was, in Thoreau, only a faint line between love and lust. In the extraordinary first line to his essay "Higher Laws," he writes: "As I came home through the woods with my string of fish, trailing my pole, it being now quite dark, I caught a glimpse of a woodchuck stealing across my path, and felt a strange thrill of savage delight, and was strongly tempted to seize and devour him raw. . . ."

He also gardened. If hunting gave him few qualms about the purity of his dedication to nature, gardening gave him plenty. He could not write off gardening as giving in to his love of wildness and adventure. To garden is to control nature. He fretted over the ethics of his efforts to grow beans, "making the yellow soil express its summer thought in bean leaves and blossoms rather than in wormwood and piper and millet grass, making the earth say beans instead of grass." He deplored the need to turn insects, cool days, "and most of all woodchucks," into enemies. He agonized over the hoeing and the weeding, the endless eradication of one of nature's species in favour of another, domesticated kind, and even despaired of turning biodiversity into monoculture. And yet he planted beans, two and a half acres of them, "so many more than I wanted." Why? he asked. Heaven knows, he shrugged, and kept hoeing.

I thought of Thoreau and his dilemma the other day when I was weeding our garden, pulling a bushel or two of invasive lamb's quarters out of the spinach patch, knowing that lamb's quarters are not only as edible as spinach, but are also closely related to it. Ditto Queen Anne's lace and carrots. Why one is a weed and the other a vegetable is a matter of arbitrary preference; pulling up lamb's quarters to plant spinach is as wasteful as killing off bison to replace them with domestic cattle. Why do we do it? Why do we eradicate the natural in favour of the unnatural? I've suggested one reason in an earlier essay: we killed bison because they were a temperate-climate species, so that we could raise domesticated subtropical cattle in their stead. But the urge to tropicalize our diet is only part of it. As Thoreau knew, nature stirs deeper desires within us.

Edward O. Wilson, the noted Harvard biologist, recognizes in human beings an innate bond with other species, an almost psychic connection he calls "biophilia," which he defines as "an urge to affiliate with other forms of life," a process so basic it appears to be programmed into the brain. "To an extent still undervalued in philosophy and religion," he writes, "our existence depends on this propensity, our spirit is woven from it, hope rises on its currents." Thoreau would have shouted his hurrahs at this sentiment; he would have seen his craving for "some kind of venison which I might devour" as the tangible expression of the same intellectual thought. There is no more direct way of becoming one with nature, he might have said with Wordsworth, than to eat it.

And yet Wilson is appalled by the way we consistently desecrate the objects of our veneration, for he has been one of the fiercest critics of the destruction of the Earth's forests and the consequent loss of species biodiversity. Confining his efforts to New World tropical rain forests, which are the most threatened ecosystems on the planet and also support the greatest number of species – a conservative estimate is

5 million species of birds, animals, and insects – he warned in 1986 that "the tropical world is clearly headed toward an extreme reduction and fragmentation of tropical forests, which will be accompanied by a massive extinction of species." By massive, he meant unimaginably large and rapid, for as forests decline in size, becoming islands of biodiversity in a surrounding sea of clear-cutting and urbanization, the number of species they can support also declines exponentially, in some cases by 50 per cent over a hundred-year span. Wilson estimates that, at the current rate of destruction, the rain forest is losing 17,500 species per year. Comparing these rates to known extinction rates in the geological past – during the time of the dinosaurs, for example – he finds that current rates of extinction due to deforestation and habitat loss are about one hundred to one thousand times the rate at which species became extinct before we began lending them a hand.

If we love nature so much, why are we so set on destroying it?

Humankind seems made up of equal parts biophilia and what nature writer Barry Lopez calls "therophobia," or fear of the beast. We are attracted to nature and yet fearful of it, and when we fear a thing we either try to tame it or destroy it. For evidence of this love-hate dichotomy we no longer need to look to the tropics, as Wilson did, for there can be no clearer case of biophilic therophobia than that of the beleaguered grizzly bear in British Columbia.

First the biophilia. One of the last things British Columbia's NDP government did before ending its mandate in June 2001 was to impose a moratorium on grizzly-bear hunting in the province while biologists made a serious effort to figure out how many bears were left and how many could be spared. And one of the first things the incoming Liberal government did was to announce a long-term, $1-million grizzly-bear reintroduction program in the North Cascade mountains of southern British Columbia. The idea was to capture five bears a year from the Chilko Lake region and relocate them to Manning Provincial Park, just north of the United States border and south of the Thompson and Nicola watersheds, roughly between Lake Okanagan and the Fraser River. The hope was to increase the bear population in that area, from at the time fewer than twenty-five, to 150 by the year 2050.

The two initiatives seemed to derive from a warm gush of nature love on the part of British Columbia's environment ministry. The Wildlife Branch finally heeded warnings from independent scientists and even its own biologists, who reported that the province's grizzlies had been declining for years as a result of habitat loss and overhunting, and that "exacerbating that decline by continuing the grizzly bear hunt is biologically irresponsible." Environmental groups such as the Raincoast Conservation Society (RCS), a watchdog organization that employs its own scientists to keep tabs on the province's wildlife habitat, had also advocated a moratorium. Most scientists suspected the number of grizzly bears in the province was nowhere near the branch's estimated thirteen thousand – the figure upon which it based its annual allowance of three hundred bear licences to the province's hunting community – and was probably closer to four thousand, although no one could really say for certain, because no actual head count had ever been made.

According to RCS's Misty MacDuffee, the government's bear-density estimates "were numbers pulled out of a hat," arrived at by calculating how many hectares of viable bear habitat remained in the province and multiplying that number by the estimated number of bears a viable hectare ought to support. But that did not take into account natural bear mortality or predation by licensed hunters, unlicensed poachers, irate farmers, encounters with logging trucks, falling trees, and the thousand other shocks that bear flesh is heir to. When a petition signed by sixty-eight independent biologists called for a five- to ten-year ceasefire on all sport hunting pending the completion of scientifically valid, long-term population studies of the province's six bioregions, the NDP government, perhaps desperate to appear to be doing something right, agreed, at least in the tentative way governments have of agreeing with anyone. It granted the scientists a three-year moratorium on hunting and told them to get counting. So far so good.

The Liberals' reintroduction program in the North Cascades was a tad more complicated. First off, reintroductions in general are dubious propositions, in that they are made necessary by our having already extirpated a species from the areas where the reintroductions are being made. Logging – or, to paraphrase Thoreau, our penchant for making the woods say boards instead of bears – had taken its toll on grizzlies in the North Cascades. Was logging to cease there? And then, of course, the idea of taking bears from somewhere else so as to introduce them to the North Cascades raised some uncomfortable questions about what was happening at Chilko Lake. Who estimated that Chilko Lake could afford to give up twenty-five grizzly bears over five years without affecting the remaining population's ability to reproduce and prosper, if there was a remaining population? Wasn't that estimate done on the theoretical bears-per-hectare model, the very formula that the scientists themselves said caused the near-extirpation in the North Cascades? Weren't we robbing Peter to pay Paul, trying to shore up our biophilia with the tattered results of our therophobia?

There is a small but influential contingent of environmentalists who say no to all reintroductions – who, in reference to the remaining wild individuals of a once prospering species, say let them go. This was the controversial position taken by the archdruid of American ecology, David Brower, when he contemplated the removal of the last six California condors from the wild in the early 1980s. Condors were meant to "live free," he said. "Let them die with dignity." The National Audubon Society, the Sierra Club, and Friends of the Earth agreed with him. Brower's point was an ethical one: condors have been on a slow decline since the end of the last Ice Age, and rounding up the last survivors and breeding them in zoos was tampering with a natural process. But the environmental groups were also making a political statement: remove the last condors from the wild, they were saying, and you remove any incentive to protect and restore the condor's habitat. And reintroducing captive-bred condors into unrestored habitat is not merely senseless, it is irresponsible.

Something of the same position was taken by opponents of the B.C. Wildlife Branch's bear reintroduction plan. No one was saying the government should let the North Cascades population die out, but many felt that any plan was fatally flawed that involved removing bears from a habitat where no reliable estimate of the bear population existed and sticking them into another habitat which had already demonstrated its inability to support bears. Brian Horesji, for example, an environmental consultant in Calgary, agreed that the North Cascades bear population was desperately in need of protection, since (like the California condor) it had "persisted marginally close to extinction for approximately a century." But he found the relocation proposal contained "major deficiencies," since it failed to place the bears' interests ahead of those of mining, forestry, farming, or even commercial recreation, the very influences that had been responsible for extirpating the bears that had once inhabited the area. The environment ministry was bending over backward to assure everyone that introducing the new bears would have almost no impact on present land uses. The woods could continue to say boards, and the earth could go on saying bears. Such a plan wouldn't work, said Horesji. It was not species reintroduction, it was "dumping," and could easily end up threatening the bear populations in both the North Cascades and Chilko Lake areas.

The flaws in the bear reintroduction plan pale beside an even more graphic illustration of our contradictory relationship with grizzlies, for in almost the same breath that the Liberal government announced its grizzly-bear program (biophilia) it also promised to lift the NDP's moratorium on bear hunting (theraphobia). "Despite widespread support for the grizzly hunt moratorium from the independent scientific commu-nity, the tourism industry, conservation organizations, First Nations, and the general public," said Chris Genovali, execu-tive director of Raincoast Conservation Society, the premier of British Columbia "has chosen to pander to an extremist minority of sport hunters who favour killing grizzlies for fun and profit." It was the kind of push-pull thing that drives environmentalists wild, giving with one hand and taking away with the other. It was equivalent in its psychological impact to George W. Bush's determination to "unsign" his country's commitment to the Kyoto agreement on global warming. In British Columbia, polls showed that the mora-torium was supported by 78 per cent of residents, 66 per cent of hunters, and 78 per cent of the people who voted Liberal in the last provincial election. Ending it seemed to make no political, let alone biological, sense.

The seventeenth-century German theologian Jakob Boehme grappled with the idea that good and evil seem to hold equal sway in regulating the affairs of humankind. "In Yes and No," he wrote, "all things exist." If you think of Yes as biophilia and No as therophobia, it makes a kind of psy-chological sense that the same administrators who said Yes to a grizzly-bear reintroduction plan, flawed though it was, could say No to a moratorium on grizzly-bear hunting. Did we want to save the bears? Did Thoreau want to grow beans? Yes and no.

Theoretically, there is no doubt that we are closely connected to nature and celebrate its presence among and within us, as E.O. Wilson observes. But Wilson also acknowledges that nature alerts a strange and primal response in us, a nexus of emotions that contributes to the complexity of our relationship with nature. This push-pull, Yes/No dichotomy often shows up in the language we use to describe what we're doing. Wilson cites as an example the connotative differences between a snake and a serpent. The former is a product of nature, a flesh-and-blood reptile, while the latter is a construct of culture, "a demonic dream-image." I see the same dichotomy in our description of food animals: we don't eat cattle, we eat beef; we raise sheep but eat mutton; we feed the pigs and eat the ham, we hunt deer but eat venison. Perhaps its our bicameral minds at work, but we respond equally and instantly to both sides of the dichotomy: when we see a writhing creature at our feet, we think snake and serpent at the same time, and are both attracted and repelled.

We see these warring elements constantly at play in our everyday dealings with the natural world. We approve of the existence of rodents, but not in our pantries or granaries. We are fascinated by ant colonies, unless they happen to be located in the foundations of our garden sheds. We appreciate wild plants, but not as much as spinach. We want to save grizzly bears and wild ducks from extinction – so that we may hunt and kill them. We are made of equal parts biophilia and therophobia, and environmentalists as well as legislators would be wise to allow for that inherent duality. For as Thoreau knew, despite our stated admiration of the wildwood, each generation "is very sure to plant corn and beans . . . as if there were a fate in it."

KILLERS IN OUR MIDST

In March 2001, an Alberta television station aired a newscast depicting Alvin Harter and his wife, Lucy, engaging in their favourite weekend pastime: killing coyotes. The camera took us along on one of the Harters' outings. It showed them spotting a coyote and letting their six hunting dogs out of their cages in the back of their truck. It followed the Harters and their dogs as they bounded across the prairie after their quarry, baying excitedly (the dogs, that is, although the Harters were fairly agitated, too). It closed in graphically on four of the dogs grabbing each of the coyote's legs while the other two chewed on her throat and belly. "Good boys," we heard Alvin Harter say off-camera. Then it was back in the truck and off to find another victim. The dogs caught four coyotes in a single afternoon. When they were too slow to kill one of them, a female in estrus, Lucy herself waded into the melee to finish the job with a hammer.

The Harters had been hunting coyotes in this fashion for many years – "It keeps us young," said the seventy-three-year-old Alvin – adding that they were welcome on most farms in the Edmonton area. "The only people who object to it are little ladies in pink hats who live in apartment buildings," he said, which isn't quite true unless the SPCA and Alberta's environment ministry are peopled entirely with little ladies in pink hats. When Al Gibson, a wildlife biologist with the ministry, deplored the Harters' hobby as "an archaic practice," Alvin replied, "I don't want to kill all the coyotes in the area, I just want to have fun."

On January 9, 2001, also in Alberta, wardens in Banff National Park shot a cougar. Earlier that day, a young woman named Frances Frost had left her parents' condominium in Canmore to do some cross-country skiing near Banff. She was thirty, a dancer and artist who, five years before, had moved to the small town just outside the park to be closer to the mountains. About twelve kilometres from the Banff townsite, she detoured along the Cascade fire road, evidently intending to hook up with the main trail to Lake Minnewanka. Just before she reached the trail, a cougar, which had been watching her approach from behind a tree, let her pass by, then left its cover, stalked her, leapt upon her from behind, and sank its fangs into the back of her neck, killing her instantly. Shortly thereafter a second skier happened by and alerted the park wardens. When they arrived with their rifles, they found the cougar still crouched over its kill.

The wanton and even gleeful destruction of coyotes and the dispatching of a rogue cougar that has developed a taste for human flesh represent two points on a continuum that defines our tolerance toward predators. In both cases, humans are killing wild animals that sometimes prey on our livestock or even on us. As the cougar incident illustrates, there are times when it seems acceptable – perhaps necessary – to kill predators. Yet, as the Harters' hobby hunting shows, there are situations when we seem to be exterminating them far too readily and for no good reason. The question, I suppose, is, When is it okay to kill an animal for doing what comes naturally?

Moral outrage aside, the Harters' coyote rampage isn't justifiable even on a practical level. Ranchers and farmers who complain about coyotes ravaging their domestic stocks simply aren't taking appropriate measures to protect their own investments. Wild carnivores work on something called (by us) the "optimum foraging theory": a predator will go after the easiest available prey, seeking to expend the least amount of energy to acquire the maximum number of calories. Under normal circumstances, about 95 per cent of a coyote's diet consists of ground squirrels, field mice, and other small rodents that are easy to catch and digest. A coyote has to eat a lot of them, but there are a lot of them around. They are easier to catch than sheep or cattle, and they can be hunted singly or in pairs, not requiring pack hunts. Coyotes rarely hunt in packs, preferring the stand-still-and-pounce method. They hunt with their ears, not their legs. At certain times of the year, lambs and newborn calves are easier to hunt than ground squirrels, but that's when a farmer's responsibility to his or her livestock comes in. Calving and lambing dates are predictable, even calculated, and at those times greater vigilance can be exercised. Pregnant animals can be corralled closer to home, where lights, loudspeakers, guard dogs, donkeys, llamas, or ostriches can be used to ward off predators. (With caution. Llamas and ostriches are used to guard sheep because they are extremely territorial and will chase off coyotes to defend the territory occupied by their sheep. Sometimes, however, they can be too territorial. I recently talked to a sheep farmer who had to get rid of his ostriches because they were killing his sheep.)

Farmers can also stop attracting coyotes to their property in the first place by calling in dead-stock removers when one of their animals dies, rather than simply dragging the carcass out into the field to rot, as most still do. A study conducted by Arlen Todd, a researcher with Alberta Environment, showed that farm carrion comprised 60 per cent of a coyote's diet in the farming regions of that province; when farmers near Westlock started burning or burying their dead stock, the coyote population in the area declined by 93 per cent. The trick is to remind the coyotes that ground squirrels really are easier prey. Coyotes are not stupid.

As for the morality of the Harters' coyote hobby, coyotes are all too often killed in inhumane ways: poison embedded in planted carcasses; gasoline poured into dens, followed by a match; a nerve gas developed during the Second World War, known as Compound 1080, which one researcher has called "the most inhumane poison ever conceived by man." An especially sadistic device is the "go-getter." Made from a piece of scented cloth tied to a detonator that's attached to a device filled with cyanide pellets, the go-getter is buried in the ground with the bait cloth poking up through the surface. When the curious coyote tugs on the cloth with its teeth, the explosion blows the cyanide into its mouth and eyes. The worst of these methods have been outlawed, but predator catchers on Ski-Doos still run down coyotes in deep snow, keeping them running until their hearts burst. Others trap a number of coyotes during the day, pour gasoline on them at night, set them on fire, then release them for the pleasure of seeing them light up the prairie like giant fireflies. None of these methods has anything to do with protecting livestock. The delight and ingenuity they reflect seem to have more to do with revenge. Or fear. All of them suggest that, in the case of coyotes at least, our tolerance for predators in our midst is pretty low.

Which brings us to cougars. We don't seem to want them around either. Despite a hundred years of concerted effort, there are more coyotes around than ever; not so with the cougar. Although cougars (also called pumas, panthers, painters, catamounts, and mountain lions) once ranged from coast to coast and from Tierra del Fuego to the Arctic Circle in twenty-six subspecies, four centuries of development and hunting have reduced them to dwelling almost exclusively between the Rockies and the West Coast, and on Vancouver

Island. Current estimates place about eight hundred of them on the eastern slopes of the Rockies (including two dozen in Banff National Park), another eight hundred on Vancouver Island, and in isolated pockets elsewhere in the B.C. interior and the Yukon.

East of the Rockies the cougar has been officially extinct since the turn of the last century – the last eastern cougar seen in Ontario was shot near Creemore in 1884 – but scat found by a trapper near Kenora, Ontario, in 1998 was subjected to layer chromatography by the Ministry of Natural Resources and declared to be "probably" cougar. MNR biologist Neil Dawson remarked that the find might signal "the return of a predator, or just . . . proof of something that's always been here." Sounds a bit like Sasquatch. There have been more recent reports of sightings elsewhere in eastern North America, however, and several of them have been confirmed – two organizations, the Eastern Cougar Network and the Eastern Cougar Foundation, keep track of such reports and try to verify them when possible. A woman north of Kansas City, Missouri, for instance, hit a cougar with her car in October 2002, and when the animal was subsequently tracked and shot by police it turned out to be a two-year-old male weighing 125 pounds, with deer hair in its intestines. Another cougar was caught by a wildlife camera in Minnesota in April 2002. No one can be sure, however, that these sightings of cougars in the wild are not of animals that have escaped from zoos or been released as pets that outgrew their cuteness.

Despite claims to the contrary by diehard cougar enthusiasts, there has been no documented attempt to reintroduce eastern cougars to their original habitat (even though elk have been reintroduced in Ontario and are already being decried as "a menace"). As wildlife biologist Adrian Forsyth has observed in his book *Mammals of the Canadian Wild*, "perhaps there is an element of fear in the reluctance to promote this feline, for pumas grow to a great size." Although the average weight for an adult male is around 150 pounds, Theodore Roosevelt once shot one that weighed 220 pounds, and the largest recorded animal, killed in Arizona in 1917, weighed 276 pounds. The cougar that killed Ms. Frost was a healthy "suboptimal" male, an adolescent, weighing 132 pounds and measuring six feet nine inches, not counting the tail.

The late Canadian nature-writer R.D. Lawrence, who once made his living hunting cougars in British Columbia, referred to the animal as the "ghost walker." He described it as a highly evolved hunter, "silent and cautious as a rule, but exceptionally noisy when moved to utter its fearful cries of love or rage." As it pads through the forest, "it makes but the merest whisper of sound, lithe and graceful and perhaps more alert than any other North American predator." Like all felines, it is nocturnal and partly arboreal, which perhaps adds to its mystique. By coincidence, I picked up Seamus Heaney's translation of the Old English poem *Beowulf* shortly after reading Lawrence's *Ghost Walker*, and was struck by the similarity between Lawrence's description of the cougar and the *Beowulf* poet's evocation of the mythical monster Grendel, "the shadow-stalker, stealthy and swift," "the terror-monger," "the dread of the land." Like the cougar, Grendel is the personification of all that is evil and terrifying in nature, even though the monster itself attacked only a single human habitation.

In modern times, cougars have assumed a similarly mythical role, instilling far more dread in us than their numbers or the frequency of attacks seem to warrant. Paul Beier, a wildlife biologist, has gathered evidence of cougar predation since the first recorded attack on a human — Phillip Tanner of Betty's Patch, Maryland, who was killed by a cougar on May 6, 1751. Beier points out that more people die each year from dog bites than from marauding cougars. Almost as many people have been killed in the past century by bison and elk as by cougars, and we are busy saving both bison and elk from extinction. And although there have been twenty-nine recorded deaths from cougar attacks since 1890, three hundred people have been killed in that time by bears.

On the other hand, as Beier also notes, cougar attacks seem to have accelerated in recent years, with more attacks between 1990 and the present (eighteen) than in the preceding ten decades. A few weeks after Ms. Frost's death, for example, a cyclist was attacked on a highway near Port Alice, on Vancouver Island, and a month later a man and his wife were attacked while camping near Rupert Arm, also on Vancouver Island. While the average number of cougar encounters in North America remains around 3.5 per year, the area around Banff experienced five in two months, three of them on a single day: ten hours before Ms. Frost was attacked, a different cougar attacked a dog in Canmore, and that same morning, also in Canmore, a cougar backed Cheryl Hyde and her pet schnauzer against a fence and was about to pounce when a neighbour intervened and chased the animal off. If there is a sense that mountain lions are coming down from the mountains and waiting for us at the gates, like Grendel, that sense is not unsupported by statistics.

The increasing frequency of these attacks has been attributed to several causes, most of them having to do with human encroachments into cougar territory. Rebecca Grambo, in her book *The Cougar*, suggests that hunters may be killing off the more passive or less intelligent cougars, causing the species to select for aggressive, clever animals. Habitat-loss and hunting may also be reducing the cougar's traditional prey species: elk and mule deer in the West, white-tailed deer and moose in the East. Logging on Vancouver Island and resort and highway development in Alberta's Bow Valley are rapidly replacing four-legged prey species with a certain noisy, brightly coloured, bipedal alternative, and cougars may be adapting accordingly.

Cougars do adapt quickly and sometimes quirkily. Researchers in British Columbia's southern Kootenay region recently set out to discover why that area's mountain caribou population suddenly began to take a nose-dive, particularly in the small area between Kootenay Pass and the U.S. border. Although caribou don't normally form a big part of the cougar's diet, the suspicion was that cougars in the Kootenay region had developed a preference for caribou meat and were killing off the herd one by one. Some wildlife managers proposed a massive thinning of the province's cougar population, a proposal that found some popular support in light of the recent attacks on humans.

After several years of study, however, the researchers found that some cougars – not all, perhaps only one or two – can take a sudden, inexplicable fancy to a certain type of meat (in this case, caribou) and stick with it even when other types are more readily available. In other words, biologists are beginning to suspect that some cougars are not opportunistic hunters all the time; at some point, for some reason, certain individuals become highly specialized, rather like my oldest daughter, who one day decided she wouldn't eat any meat other than fish, and didn't for two years. The same tendency in cougars, if it turns out to be true, would call for a modification of the optimum-foraging theory. It may be that specializing in a particular prey species, even when that species is less numerous and harder to catch than others, requires less energy than opportunistic hunting does. For a carnivore, honing its skills and sharpening its knowledge of a single prey species may be a surer way to guarantee a regular food supply than simply waiting around for something easier to kill. In human terms, it would be like a group of hunter-gatherers deciding they'd rather build a community on the coast and become wise in the ways of the sea and its denizens, than continue wandering around in the forest hoping to come upon a solitary deer. Every time my daughter went into a restaurant she knew in advance what she was going to order, and that may have reduced any stress she might have felt about going into restaurants. The wolves in Banff National Park exhibit the same trend. The eleven-member pack near the Banff townsite seems to prefer to hunt the elusive and aggressive mountain sheep rather than the docile and more abundant mule deer that have concentrated in the region. By studying the quirks of these highly evolved predators, biologists have been afforded a rare and precious glimpse into the way evolution works. They are also learning that if we want to control predation on a particular species, we may be able to do so by eliminating one or two rogue individuals that have shown a taste for that species instead of engaging in a general blitzkrieg on the whole population of carnivores.

The lesson can easily be applied to coyotes. Farmers who suffer hefty losses to coyote predation may be able to solve their problem by live-trapping one or two problem animals rather than by exterminating every coyote in sight. Similar reasoning can be shared with city-dwellers concerned about predators migrating from the wild. In Vancouver, cats and dogs have been disappearing at night as more coyotes move down the Fraser Valley to the fringes of the city. The coyotes may have started killing pets to eliminate the competition for mice and squirrels, and as a result developed a taste for domestic animals raised on Miss Mew and Puppy Chow. (My theory is that the high cereal content in commercial pet food makes domestic cats and dogs taste like herbivores.) Grieving pet owners have called upon the government to destroy the coyotes, but a band of more nature-minded citizens started a project known as Co-existing with Coyotes through Communication. Their aim is to increase understanding of the dynamics of coyote/human interactions, and to lobby the government to refuse to sanction the killing of wild predators. Perhaps pet owners, too, could be encouraged to protect their animals by keeping them indoors at night, when coyotes are on the prowl, rather than by urging mass extermination of the suburban intruders.

In Edmonton, a similar group, named Voice for Animals, has begun lobbying the provincial government in direct response to the newscast of the Harters' weekend hobby hunting. "Our goal is to have changes made to the law to prevent the hunting and killing of coyotes by dogs," says volunteer Tove Reece. "It's going to take some education of the public before we can attempt that." As the Harters have shown, predator hunting is almost always a blood sport – think of fox hunting, which was banned in England in 2004 – no matter how vociferously it is defended as predator or vermin control.

The cases of coyotes and cougars, it would seem, are not so far apart on the continuum of tolerance. We could solve any real problems we have with these predators by dealing with them on an individual basis. But we don't: we demonize the whole species instead. In the end, both animals illustrate the need for a better understanding of the role of the wild in our lives – even our urban lives. As Roslyn Cassells, a commissioner on the Vancouver Park Board (which partially funds the B.C. coyote project), puts it: "Coyotes are here to stay. We must learn how to live with them, and respect their ways." Following the death of Ms. Frost, the Banff Springs Hotel put a sign on its check-in desk warning guests that "the wilderness is just a few feet away in any direction" – a statement that applies everywhere and is always true.

WAR AND PEACE IN BIRDLAND

I n 1908, Anatole France wrote a satirical novel called *Penguin Island* that pretty well summed up the question of aggressiveness in birds. The novel recounts that at some point during the impenetrably Dark Ages a Welsh monk named St. Maêl sailed to a land north of Iceland, where he discovered a large population of penguins. Being half dead and delirious from cold and starvation, the holy man mistook the penguins for a species of diminutive but formally dressed human beings, and baptized them. Because God is morally bound to accept any baptized creature into His kingdom, St. Maêl's rash act plunged Him into a theological quandary; what could He do with a bunch of baptized birds? After much debate among the saints, He decided to solve the dilemma by turning the penguins into humans. It was a case of God playing at being a bioengineer.

Like many satires, France's novel purports to be non-fictional history, in this case of the Penguins from the date of their transformation to the present. He seems to have known a thing or two about bird as well as human behaviour. At first the penguins-turned-humans were a decidedly disagreeable and aggressive race. "For thirteen centuries the Penguins made war upon all the peoples of the world," France writes, "with constant ardour but fluxuating fortunes. Then for a number of years they tired of what they had so long loved and displayed a marked preference for peace, which they expressed with dignity and, it seemed, deep sincerity."

War and Peace in Birdland? Well, why not? We are accustomed to thinking of birds as aggressive creatures: Aves, as Alfred Hitchcock knew, can be an extremely bellicose class. Birds are often their own worst enemies. They pluck one another from the air (I once watched a peregrine falcon in downtown Milwaukee feeding on its favourite prey species: a blue jay). They shove one another from nests to get at eggs and hatchlings (a heron hatchling's chief predator is the black vulture). They bully their own kind from feeders (blue jays again, but they're not alone) and roadkills (ravens and turkey vultures). It isn't called a pecking order for nothing. In short, they behave a lot like humans. Anatole France may have been on to something. The ancient Chinese thought that human beings were descended from birds, and that we inherited many of our aggressive tendencies from them. And Aristophanes (448–388 BC), in his play *The Birds*, asserted that messages from the Gods to Earth were routinely intercepted, reinterpreted, and often rerouted in Cloudcuckooland, a Purgatory-like realm located between us and the empyreal regions and administered solely by birds, who were cruel, self-serving despots. Like humans.

When viewed in this light, it's not the War half of *War and Peace in Birdland* that surprises, but the Peace. Even among penguins, which seem on the surface to be so, well, comfortable with one another, so dons-at-the-cocktail-hourish in their shabby tuxedoes and their short flipper-like wings, just the right length for holding napkins and sherry glasses, even among so dumpy a set we suspect smouldering resentment. Yes, we've seen *March of the Penguins*, in which researchers walk amidst hordes of unconcerned Emperor penguins in Antarctica, hands unprotected by gloves, pant legs intact, but we don't believe it. There must be stories of being attacked by angry penguins defending their young, or their mates, or their dinners.

As it turns out, there are. Penguins are highly aggressive animals. Of the seventeen species of Spheniscidae, only the largest, the Emperor penguin, weighing in at thirty-nine kilograms, is considered even remotely amiable, possibly because they do not have regular nesting sites and so do not defend their territories. But how amiable is amiable? In the documentary, Emperor penguins push other Emperor penguins into the water to see if there are any killer whales about. Female Magellanic penguins fight like fishwives, and in Humboldt penguins a ceasefire of one minute between periods of fierce aggression is considered a United Nations triumph. Chinstrap penguins are particularly aggressive, especially during nesting season, when they steal nest sites and even nest material from one another. Male Adelie penguins are even more belligerent at such times: females lay their two eggs in little gravel nests, up to two hundred nests in a fifty-square-metre area, then swim thirty kilometres out to sea to fatten up on krill, leaving the males behind to incubate the eggs. The males sit on the eggs for up to a month without eating, which would make anyone a little testy. Carol Vleck, a zoologist with Iowa State University, upon analyzing blood samples from the necks of Adelie penguins on Antarctica's Torgersen Island, found extraordinarily high testosterone levels in males at breeding time (October), which, she hypothesized, makes them behave in thoroughly penguin-like ways. Unmated males steal eggs and chicks from other nests, and even successful males peck neighbouring males nearly to death to gain a few extra centimetres of space for their own offspring. They protect eggs and fledglings from marauding skuas with a ferocity that Vleck found alarming.

In short, even donnish penguins can act like Vikings when the spirit, or the hormones, moves them. Other seemingly unaggressive birds also have their dark side. Doves, for example. We refer to peace-mongers as "doves" (as opposed to war-mongering "hawks"), but that distinction amuses ornithologists, since doves are anything but peaceful. Konrad Lorenz once put a pair of doves in a box and watched the female tear the male apart. "Like an eagle on his prey," he wrote in *On Aggression*, "stood the second harbinger of peace!" It may help to recall that it was the surrealist poet

Louis Aragon who chose Picasso's drawing of a dove to be the international symbol of peace – the figure was officially adopted by the World Peace Congress on April 19, 1949. Picasso himself was aware of the irony: "There's no crueller bird," he said of the doves he had kept in his studio as models for the drawing: "They pecked a poor little pigeon to death because they didn't like it. How's that for a symbol of Peace?"

That other famous symbol of peace, the three-pronged fork enclosed in a circle, which those of us who remember the sixties saw emblazoned everywhere, molded into medallions and belt buckles and spray-painted on the flanks of missiles, was taken from a Hopi design that, according to Peter Matthiessen, was based on the footprint of the sandhill crane. To the Hopi (as to the Chinese and Japanese), the crane is the sublime embodiment of tranquility. But as Matthiessen notes in his book *The Birds of Heaven*, cranes are fiercely aggressive; a mating pair will defend a huge area from other nesters, and when its young are hatched will chase away not only other cranes, but also ducks and grebes and even aquatic mammals, anything that may compete with its own offspring for food. Cranes as symbols of Peace? Ask a mallard.

Anatole France was indeed not far from the truth when he described his birds as alternating between long periods of aggressive behaviour and intervening episodes of tranquility. Neither was he far off when he assigned the pattern to penguins. Vleck noted that when testosterone levels crashed, as they did when the chicks hatched and the females returned from their seaside holidays, male Adelie penguins turned from snappy aggressors to docile *pères-de-familles* almost overnight.

This isn't quite what Anatole France meant. He wrote about peaceful intervals that lasted years, not weeks, but since he was writing about great auks, not true penguins – which, of course, are not found in the north – he may have known something we don't. There is, however, no reason to suppose great auks behaved much differently from penguins; the similarity between the two certainly struck Sir Francis Drake, who transferred the name "penguin" to the southern bird. In another northern species, ptarmigan, such a see-saw pattern between warring and peacing has indeed been observed by researchers. I spent much of one winter in the Yukon, and a good part of my time there in the pleasant company of Dave Mossop, an ornithologist originally from Manitoba who was once the Yukon government's Senior Biologist and is currently a biology instructor at Yukon College in Whitehorse, where he heads the Biodiversity Assessment and Monitoring Project. The biodiversity he mostly assesses and monitors is that of birds, particularly peregrine falcons and gyrfalcons, two dangerously threatened raptorial species not unassociated with aggressive behaviour. While he was with the Yukon's Department of Renewable Resources, however, he did a great deal of work on ptarmigans, and raised some intriguing questions about the nature and consequences of pacificity and aggression within a single species.

Like the dove, the willow ptarmigan (*Largos largos*), a member of the grouse family, is to all appearances a gentle bird. Cryptically a light, speckled brown in summer, when it blends in with its lowland habitat, in the winter its plumage is as white as the land is, the purest, starkest white imaginable.

The most widely distributed game bird in the Arctic, it appears in Native mythology as a demure, virginal creature much admired by its more predatory neighbours, to wit, crows and ravens. Alaskan Native tradition has it that Willow Grouse Woman was so beautiful that "all the boys were after her," but, as is often the case with beautiful virgins, "she didn't want to marry anybody." When Crow, dressed in false finery, courted her, she refused him out of hand, and so he carried her off and "dirtied" her anyway, which accounts for the few parts of her that aren't white: her head and neck in springtime, her coal-black tailfeathers in winter. In another version, Crow actually kills Willow Grouse Woman with a spear, and a willow ptarmigan flies out of her body. In both versions, Crow is the aggressor, and Willow Grouse Woman the victim of outside forces that bring about her demise.

Dave Mossop knows different. His research was aimed at explaining why willow ptarmigan populations crashed and boomed so spectacularly every nine to eleven years. Each decade or so, the number of mating pairs per square mile soars to as high as eighty and then plummets to fewer than twenty within a few years, decreasing at the astonishing rate of about 50 per cent per year. No one knew why that should be so, but then no one had studied ptarmigan populations in winter. Winter, in the places inhabited by willow ptarmigan, does not lend itself to sedentary birdwatching. Think of Sam McGee pausing on the marge of Lake Lebarge at fifty below to count snow geese. Before Dave came along, the bird had been studied only in summer, when bonding pairs were already formed and breeding strategies set. Dave decided to watch what they did during the winter, when the birds' chief concern was not breeding but survival, and to chart their transition into spring. He wanted to see if the birds died during the winter, and if so of what, or whether the crash resulted from some reproductive failure later on – fewer eggs, unhatched eggs, unfledged chicks. He wasn't trying to prove that over-hunting (a winter activity) caused the dramatic decline in population, although he was accused by the territory's hunting lobby of leaning in that unpopular direction ("Every time you write a report," his supervisor once told him, "you cause trouble." "Which was true," he cheerfully admits), but rather to see which, if any, of the three usual suspects – predation, food shortages, or extreme weather – was responsible for the crash.

For his study he chose the Chilkat Pass, a high, spectacular notch in the Coast Mountains on the road between Haines Junction, Yukon, and the seaport of Haines, Alaska. Because of the peculiar nature of the border in that region, his study area actually fell in a northern blip of British Columbia, an accident of geography that caused him interminable paperwork with Victoria.

But it was worth the effort, he says, because the Chilkat Pass is ideal ptarmigan habitat: flat, wet, covered with arctic willow – a beautiful, wispy shrub and the willow ptarmigan's year-round food of choice – and under deep snow during the winter. Ptarmigan spend most of the dark winter days buried in "snow roosts," burrows they make into the snow, a new one every day, emerging for short periods in the early morning and late evening to browse on willow buds. Although spread out into fiercely defended breeding territories in summer, from October to April they flock amiably together, probably for protection from predators, studies having shown that for prey species there really is safety in numbers.

Dave calls his kilometre-long stretch of the Chilkat Pass "the holy land." When he drives into it after a long absence, as we did in November, he rolls down the window of his ancient four-by-four, sticks his head out into the wind, takes a deep breath, and smiles. "The air is magical up here," he said. "Somehow, it seems filled with light."

Most of the Yukon in November is not filled with light. The sun appears low in the sky for an hour or two around noon, looking like an old quarter at the bottom of a bowl of thin soup. But Dave was not speaking metaphorically: the evening did seem to linger in the pass longer than in the valleys we had come through to reach it, and to be charged with a luminosity that allowed us to find the research cabin, set though it was some distance back from the road, without the aid of a flashlight. As I shovelled the door free of snow, Dave brought the boxes of supplies in from the truck, and once inside – a wild clutter of work table, two chairs, two bunks, and a bookshelf over the door – Dave lit the small wood stove while I brought in pots of snow to melt for tea. Within half an hour the cabin was so hot we had to open the door. Dave sat back with his feet up on a food locker in his best Sam McGee manner and read the logbook he'd left on the table for visitors to sign. There were a dozen recent entries. Over the thirty years since he built the cabin, it has become well known among backpackers and cyclists, who use it so regularly that it is mentioned as a refuge, to Dave's immense satisfaction, in international hiking and cycling guides. It has also been used by ptarmigan hunters and, on at least one occasion, by a group of gyrfalcon poachers, who blasted a hole in one wall with a shotgun so that they could sit comfortably inside drinking beer and work the string that released the pole-mounted falcon gin outside. Dave was decidedly not pleased about that.

On the subject of ptarmigans, however, and the mystery of their waxing and waning population (they were then at the nadir of their density, with perhaps fewer than thirty birds in the study area, mostly males), Dave's brow furrowed and the old questions returned. His study had determined that the population crashes were attributable to none of the conventional causes. Not weather, since the decline coincided with an unusual warm spell in the region. Not predation, either. A few hunters came to the area, but not many. Foxes weren't eating an inordinate number of willow ptarmigan; in fact, in five years Dave never saw a fox catch a ptarmigan. Gyrfalcons would catch a few, but not enough to precipitate a crash, and they almost always caught yearlings. And golden eagles, although normally more skilful predators than bald eagles, were in this case even less successful than gyrfalcons, because they were larger and slower and could not follow the ptarmigan into the willows. And finally, not food shortage, since the willow bushes were full of tender shoots and buds, and in the deepest part of winter, when most of the birds migrated to a lowland patch of denser willows, there were still plenty of unbrowsed buds in the area they had left.

"So it wasn't predation, and it wasn't weather, and it wasn't starvation," Dave summarized. "Something was driving the crash, but what was it?"

In the course of the study and during many subsequent sojourns at the cabin, Dave discovered several interesting social behaviours in ptarmigan that began to lead him toward a possible answer. In the early spring, the males returned to the breeding area and quickly defined their territories, each male marking off a plot of land for itself and fiercely attacking any other male that wandered into it. After two weeks of this frenzied defensive activity, the females returned and chose their mates, the preferred males invariably being those with the biggest territories, and hence the most actively aggressive, testosterone-charged representatives of their species. If God had turned these birds into humans, you'd be able to recognize them by the size of their lawn tractors.

Dave called this main group of male ptarmigans "the fighters." But there was another group of males that seemed less interested in fighting for territory, which appeared to find fulfilment in a life of carefree browsing and roosting in a large patch of willows across the road from the fighters. This second group never fought amongst themselves and never mated, even though there were females among them. They just hung out in the trees. In his scientific papers Dave called this group the "waiting flock," but in conversation he calls them "the lovers." Lovers can become fighters. When one of the aggressive males from the fighter group was killed by a predator, a male from the waiting flock would happily take its place. And sometimes a female from the waiting flock would fly over to the breeding ground and fill in for a fighter female who was incubating a clutch of eggs. But generally the waiting flock was content to just wait.

Dave was puzzled by this apparent split personality in ptarmigan, but gradually he began to see some sense in it. The crash-and-boom cycle, for instance, coincided nicely with the alternating dominance of fighters and lovers in the flock. It happens, he said, that the fighters in the main study area did so much fighting, defended their breeding territories so vigorously, that they became weak with exhaustion. "When we were capturing the breeding males for banding," he said, "sometimes a male would be so emaciated and scrawny from fighting that we decided not to capture it for fear of killing it." The waiting males across the road, on the other hand, were always in fine fettle; they maintained a healthy weight, got along with their neighbours, kept their feathers unruffled. They seemed, in Darwinian terms, fitter than the fighters. After a while, it dawned on Dave what they were waiting for.

What he realized was that the ptarmigan population begins to crash at exactly the point at which the fighting males dominate the breeding area. You might think it would be the other way around – that the population would begin to weaken when the lovers took over. Lovers are, after all, compact with lunatics and poets. But when fighters dominate, there are the most territories to defend, and the territories are the largest. The fighting males have to work hardest to define and defend their plots, and that makes them less effective breeders.

"I don't know what the mechanism is," Dave said. "I don't know if the males are too exhausted to tread the females, or whether they get so emaciated that the eggs they inseminate don't hatch, or produce inferior hatchlings. But what I do know is that just when the main flock is at its most aggressive, that's when the decline begins. There is something about aggressive behaviour that precipitates the crash. And when the aggressive males begin to disappear, their places are taken by the lovers and the population slowly begins to build up again."

Whatever the mechanism, the message is clear: warlike behaviour may be inimical to reproductive success. Recent bird studies have suggested similar conclusions: investigations of red-winged blackbird populations, for example, show that although a male red-winged blackbird spends enormous amounts of energy defending a large harem, as many as eight chicks out of ten are fathered by outside males, who sneak in to impregnate neglected females while the alpha is off defending others in a different part of his territory. "What," Dave asks, "is the point of aggressive territoriality if not to ensure the passing of one's own genes on to the next generation?"

It may be that aggressive male behaviour serves only to distract predators from the nesting area, so that the females can take advantage of the lull to mate with opportunistic rival suitors. While the hawk's away, the doves will play, as returning soldiers from various wars throughout history have discovered. Wouldn't it be a boon to doves if the true point of hawkishness were to provide opportunities for the non-aggressors among us to reproduce?

Anatole France would no doubt agree: in 1888, he began a lengthy love affair with his patroness, Madame de Caillavet, which led to his own divorce in 1893. Madame de Caillavet was the wife of the Marquis de Flers Robert et Arman de Caillavet, who was also a writer. One of de Caillavet's plays, *L'amour veille* (*Love Watches*), written in 1907, portrays a young noblewoman who marries an aggressive count despite the fact that she really loves Ernest Augarde, "a harmless bookworm." The play was possibly meant as de Caillavet's revenge on Anatole France, but it is a classic willow ptarmigan tale, and as such an ironically cheering one for bookworms.

SOMETHING FEARFUL
THIS WAY COMES

I live in Southeastern Ontario, a part of the country that in 2001 and 2002 endured the hottest, driest summers anyone could remember, summers that invited favourable comparison with the Prairie Dustbowl of the 1930s. Temperatures of 38 degrees Celsius and virtually no rain from mid-June to late-August. Plants with shallow root systems, like corn, stood brown and stunted in the fields, tassels drooping, kernels shrivelled. Hay simply stopped growing after the spring cutting. Though it remained eerily green, it did not bend or ripple in the hot wind; it stood crisp and rigid, like a desert army on the verge of collapse. In our garden I dug a hole two feet deep for the asparagus and still came up with soil as dry and lifeless as sand. No amount of watering seemed to make a difference. The potato plants died a premature death despite heavy watering, apples fell from the trees in July the first year, didn't form at all the second, and the sunflowers never flowered – too much sun. Tentatively, not wanting to sound unduly alarmist, we began to refer to "the drought."

Of all the plagues that can be visited upon a culture, drought must surely be the most dispiriting. It is not a super-abundance of something, like too many locusts or frogs. It is not a disease, like the murrain of beasts that Moses called down upon the Egyptians. Drought is a lack, an absence, a failure of nature. When the Children of Israel finally left Egypt, Moses led them through "a terrible wilderness, wherein were fiery serpents and scorpions, and drought, where there was no water," and, except for the scorpions, that's pretty much what it was like here that summer. But the Children of Israel were just passing through, on their way to the promised land; what farmers around here, and farther east, and in Saskatchewan and southern Alberta and the American Midwest, experi-enced was the realization that this *was* their promised land, and it had betrayed them. There was no going back, or moving on. Drought, a caprice of nature, a fluke of El Niño, can turn promises to fiery serpents practically overnight.

But we may not want all promises to be fulfilled. Just before the drought set in, newspapers were promising us several really nasty diseases that had escaped from the tropics, where they belonged, and were reportedly making their way north, enabled by rising average temperatures attrib-uted to global warming. Among these was the dreaded West

Nile virus, a deadly flavivirus first isolated in the West Nile province of Uganda in 1937, where it was the cause of an outbreak of encephalitis (inflammation of the brain). It quickly spread north throughout Egypt and Israel, where it was associated with meningoencephalitis (spine and brain). Then in 1960 it was found to be affecting horses in France, and by 1999 it had turned up in North America. That year sixty-two people in New York City were diagnosed with West Nile virus, and fourteen of them died. A survey conducted by the city deduced that 2.6 per cent of all the residents of Queens had been infected, though most experienced nothing worse than mild flu-like symptoms. Only those with weakened immune systems became seriously ill. The next year it spread into New Jersey and Maryland: seventy-nine cases were reported, two more people died.

In Canada, mild panic set in. In 2000, it was in upstate New York, waiting to cross the Niagara River. The next year it was in Ohio – moving west. Not for the first time, Canadians thought about closing the border; also not for the first time, Canadians realized that, when it came to disease control, there was no border to close.

There have been precedents. We at the top of the continent occasionally feel trapped by the United States, shoved up against the Arctic wall and held there while our exuberant and infectious neighbour coughs on us, rather like being pressed into the corner of an elevator by a large friend with a very bad cold. Raccoon rabies, for example, came to us from New England by means of infected raccoons hitching rides across the Niagara River in freight cars and camper trailers. In the 1970s, there was a serious outbreak of St. Louis encephalitis in southwestern Ontario, in the Windsor area, during which a number of people died. And remember the hantavirus scare of the early 1990s, when virus-laden deer mice from the American Southwest emigrated to Canada, carrying in their saliva the germs of a disease so deadly and virulent it could fill one's lungs with plasma in a matter of days? We watched the border fearfully, wondering how on Earth we could prevent deer mice from crossing it. Eventually eight people in Canada caught Four Corners Disease, as it was labelled, and died from it.

West Nile virus is even more perfidious than rabies and Four Corners Disease, in that, like St. Louis encephalitis (to which it is very closely related), it is transmitted to humans by mosquitoes. One can, without being extremely diligent, avoid swallowing deer-mouse saliva or being chomped by a mad raccoon. But who among us has never once been siphoned by the most common mosquito in Canada, the ubiquitous *Culex pipiens*? (Note the verb: mosquitoes do not bite, as black flies and deer flies do; they siphon. Black flies and deer flies, like vampire bats, make little pools of blood in our skin and then lap it up; mosquitoes inject their long, six-probed probosci into us, inject a kind of blood-thinning serum into our bloodstreams, and suck their fill – about a millionth of a gallon per fill-up.) *Culex pipiens* is an effective West Nile vector because of its omnivorous sucking habits. There are seventy-two species of mosquito in Canada; some feed only on amphibians, others on mammals. There are mosquitoes that feed only on birds – one species siphons solely from loons. But *Culex pipiens* takes its blood meals where it can find them, from a variety of hosts, including humans and birds. And it is in the blood of migrating or wind-blown birds that West Nile virus seems to have made its way from Uganda to two other continents, including our own.

Here is how it works: a mosquito siphons blood from an infected human in Uganda and itself becomes infected with West Nile virus. The same mosquito then takes a second blood meal from a bird that winters on the banks of the West Nile River and summers in, let us say, France. When the mosquito injects its blood-thinning serum into the bird preparatory to siphoning its meal, that bird becomes infected with West Nile and may eventually die, but meanwhile it returns to France in the spring and is siphoned by a mosquito. That same mosquito then takes a second blood meal from a horse, and the horse comes down with West Nile. No one knows how that cycle was transferred to North America, but every year many birds, in the course of their annual migrations from Africa to Europe, are caught in prevailing winds and blown over to our side of the Atlantic. Birders call them accidentals. However it happened, sometime in the spring of 1999 an infected foreign bird, possibly a member of the crow family, landed in a tree growing in Brooklyn and was siphoned by a mosquito.

I say crow because crows and other corvids, including jays, seem to be more susceptible to West Nile than other bird species; although many birds can carry the virus, only crows and their kin seem to die from it. In fact, by the end of 2000, more than 90 per cent of the crow population of New York had been wiped out by the virus. Before all the crows died, however, mosquitoes managed to transfer the virus to other species of birds as well as to humans. The most likely reservoir of West Nile now – the new Trojan Horse, if you will – is the diminutive, sooty-throated house sparrow, by far one of the most common birds in any city in North America, and also ubiquitous in rural regions. The irony here is that the house sparrow, formerly called the English sparrow, is itself a Europe-to-Africa migratory species introduced into North America – in New York in the 1880s – by well-intentioned but ornithologically misguided people who thought English sparrows lived on insects and hence would be a boon to American agriculture. In fact they live almost exclusively on seeds, and hence have become one of the greatest agricultural pests on the continent, an object lesson, if we needed another one, in the dangers of tampering with natural phenomena when we don't know enough about them. And we can never know enough.

But back to defending the border. If we could not keep deer mice or even raccoons from crossing the border, how could we hope to prevent migrating sparrows and crows from gaining illegal entry? Obviously, we could not. Instead, in the summer of 1999, a number of provincial health ministries organized, along the Canada–U.S. border, from Alberta through to the Maritimes, a series of Proximate Early Warning stations reminiscent of the old Distant Early Warning (DEW-Line) military installations the United States maintained in the 1960s in Canada's Far North. These PEW-Line sites were intended to warn us as soon as they detected any West Nile virus. They were staffed by chickens, called Sentinel Chickens. (Using this militaristic nomenclature, the traditional canary-in-the-mineshaft would be called a Sentinel Canary.) There were six hundred Sentinel Chickens in all, 180 of them in six sites in Ontario. All the chickens had to do was perch quietly in their cages and allow themselves to be siphoned by *Culex pipiens* mosquitoes. Once a week for four months, provincial health officials took blood samples from each chicken and sent them to local health labs, where impression smears of their blood were made and sent to the National Biology Laboratory in Winnipeg for analysis. That first year, none of the blood smears from PEW-Line Sentinel Chickens tested positive for West Nile.

This was not as reassuring as it might have been. Chickens, it seems, make lousy sentinels. "They're too slow," explains Chuck Le Ber of the Ontario Ministry of Health. Chickens pressed into similar service in the United States, he says, continued to yield negative results while crows and blue jays and mosquitoes and even people around them were dropping like flies with West Nile virus–induced encephalitis. It is a mystery why chickens were selected for such an important mission in the first place. Perhaps the ministry was leery of capturing wild birds and keeping them in cages, thereby breaking one of its own laws. Perhaps it would have cost money to contract someone to trap them.

In any case, the Sentinel Chicken program was scrapped in most regions in favour of a strategy that can be termed the Dead-Bird Defense Initiative. Ordinary citizens along the border were encouraged to look for suspiciously dead birds

– not car- or window- or baseball-struck birds, or birds caught by cats, but birds, especially corvids, whose deaths seemed unexplainable. Birds falling suddenly from the air, for instance, or discovered on the lawn without a mark on them. Birds whose deaths, had they been human, would require an inquest and an autopsy. Anyone coming across such a dead bird was urged to notify the nearest health unit, which then sent an investigator to the scene, and if the bird was indeed a candidate for an inquest, it was sent to the National Laboratory in Winnipeg, where it received an autopsy. By mid-August 2001, in Ontario alone, more than 1,500 dead birds were reported, collected, and examined. The species breakdown is instructive: 822 American crows; 399 blue jays; 13 ravens; 328 other species. Organ samples from about half of them were sent on to Winnipeg. Despite dire predictions from the newspapers – in June the *Globe and Mail* proclaimed "West Nile at your doorstep," warning that the virus "is almost certain to spread into Canada over the next few months" – not a single dead bird was infected with West Nile virus.

To what did we owe this miraculous stay of execution? Well, not to put too fine a point on it, it was the drought. *Culex pipiens* mosquitoes need stagnant water in which to lay their eggs. They are what entomologists call "tree-cavity breeders": they oviposit in water that pools in tree holes and things that resemble tree holes, such as automobile tires, tin cans, swimming-pool covers, and blocked eavestroughs. This makes them ideally adapted to city life. But when there is little rain, there are fewer places for *Culex pipiens* to lay its eggs. Health Canada confirmed that in the dozens of mosquito pools it monitored that summer – poised to spray if any of the dead birds tested positive for West Nile – mosquito numbers were down. Reduced mosquito populations meant a lower chance that the West Nile virus would be transferred from a bird to a human.

This was, of course, only a temporary reprieve, for if anything definitive can be said about the weather during this time of global climate change, it's that it is inconsistent. Traditional prairie wisdom has it that droughts come in four-year cycles, but global warming has no respect for traditional wisdom. In 2002, while the East was as dry as the Sahara, the Prairie provinces had more rain than they knew what to do with. Alberta's Sentinel Chickens began testing positive. Moreover, studies showed that the virus was adapting more rapidly than our defenses against it. Birds no longer needed mosquitoes to pass West Nile virus on to other birds, as was formerly the case. In one experiment, researchers placed fifteen crows in a mosquito-free aviary; nine were infected with West Nile, but the other six were not. The infected crows died after a few days, and then five of the remaining crows died, having contracted West Nile without the aid of mosquitoes: it's thought that the virus was passed by means of feces and saliva in the birds' drinking water. But it might also have just travelled through the air.

West Nile is now officially listed as an endemic disease in North America; that is, one that cannot now be eradicated. To date, only two of the contiguous American states – Oregon and Washington – have reported no cases of West Nile. In California in 2005, eighteen people and 2,986 birds died of

West Nile virus. The United States's Centers for Disease Control and Prevention estimates that every case of meningoencephalitis reported (and not all are reported) represents 150 cases of West Nile, which means that the 2,581 cases reported in 2005 represent nearly 300,000 cases of West Nile. Most of those don't get much beyond the runny-nose stage, but since 2001, 855 have resulted in death. The disease is also no longer carried solely by *Culex* mosquitoes: two introduced species — *Aedes albopictus* (the Asian tiger mosquito) and *Orchlerotatus japonicus* (the Asian bush mosquito) — have now been identified as vectors. As global warming continues, these tropical mosquitoes and the virus they carry are breeding steadily north: in April 2003, West Nile was found to have over-wintered in a mosquito population in central Minnesota. That same year, the USDA's Animal and Plant Health Inspection Service reported 5,181 cases of equine West Nile: humans and birds are not the virus's only hosts.

Perhaps most alarming is the suspicion, reported in a recent issue of *Harper's* magazine, that in some of those three thousand cases, the virus may have been transmitted by means of organ transplants and blood transfusions. A virus can't get much more endemic than that.

Along with the many other pathogens and their vectors brought here from the tropics — SARS, avian influenza, tiger and bush mosquitoes, Asian longhorned beetles — it seems that the West Nile virus will always be among us. We could think of it as the flip side of globalization: in a warming world in which everyone wears the same T-shirt and drinks the same cola, it is inevitable that we will also share the same diseases.

BLOOD RELATIONS

At a party recently, I found myself talking to a young man who, in his mid-twenties, was stalled somewhere between first- and second-year university. He kept registering for courses, he said, dallying in astronomy, oceanography, computer programming, not turning up for classes, not really knowing what he wanted to study. Although he professed to being rather aimlessly adrift, he didn't seem much bothered by the prospect of remaining rudderless. I asked him why he kept taking courses in subjects that did not interest him.

"Because as long as I keep going to school," he said, mistaking candour for honesty, "my parents will send me enough money to live on."

I couldn't help making the observation that by living off someone else's generosity he ran the risk of being branded a parasite, to which he replied, straightforwardly enough:

"What's wrong with being a parasite?"

It was a fair question, and when directed to someone interested in natural history, an intriguing one. What, in the natural scheme of things, *is* wrong with being a parasite?

Throughout modern social history the term "parasite" has been applied pejoratively, as the wording of my observation – "branded a parasite" – suggests. The OED notes that the word is used "always with opprobrious application," and lists such synonyms as "smell-feasts," "hangers-on," and "toad-eaters," from which we derive the word *toady*. In Roman times, parasites hung around noblemen's houses, exchanging gossip and flattery and errand-running for meals, favours, and other forms of social acceptance. In Shakespeare's *Timon of Athens*, Timon rebukes his sycophantic courtiers by calling them "mouth-friends . . . most smiling, smooth, detested parasites," and you can't get much more rebuked than that.

But if we go back a little farther in time, the issue becomes less clear. Our word *parasite* comes from the Greek *parasitos*, from two words meaning "beside" and "feeding," literally a person who eats with another, presumably at the other's table. A *parasitos* was originally the guest of a high priest who was obliged to invite a commoner to share the sacrificial meal, usually of barley, prepared for the annual public feast. According to Plutarch, a parasite was thus one who had been made "venerable and sacred," selected as he was from *hoi poloi* to ensure that the feast be a public event, not a private banquet for the higher-ups, since it was *hoi poloi* who provided the hecteum of barley for the feast in the first place.

Nothing wrong with that, you might say. But it's not hard to see how the aristos might come to view the *parasitos* as an imposition, perhaps even a threat, someone who didn't know which fork to use before pocketing it. The parasite himself might feel this slight, might eventually consider himself as one whose presence at the feast was tolerated by his host but not really welcomed. He might believe it necessary to secure his continued position among his betters by obsequiousness and flattery. However it came about, it wasn't long before being called a parasite slipped a person disastrously down the social scale, from one chosen to eat at the table of the gods, to a social pariah who has, in the words of E.M. Forster, "no function either in a warring or peaceful world."

The question my young friend was asking, then, was a good one: Does a parasite have a function? And the answer, not surprisingly for a class of organism that has survived on this planet much, much longer than we have, is, Yes, of course it does.

There are far more species of parasites on Earth than there are of vertebrates, and parasites survive like nobody's business. Fossilized Cambrian trilobites show evidence of having been attacked by parasites. Tetrabothrid tapeworms infested ichthyosaurs 100 million years ago, and are still around today, though ichthyosaurs are long gone. There are 4,500 species of mammals on Earth, and five thousand species of tapeworms.

When asked what his scientific studies had taught him about God, the eminent Victorian biologist J.B.S. Haldane replied, "He seems to have had an inordinate fondness for beetles." A modern entomologist could state, with equal accuracy, that God seems to have had an even greater fondness for parasites. The vast majority of all organisms on Earth are parasitic.

In nature, a parasite is a creature that makes its living by drawing sustenance from another creature, called – in a nice echo of its original meaning – its host. Not a "beside feeder," then, but an "inside feeder" if it's an endoparasite, like a tapeworm, or an "outside feeder" if an ectoparasite, like a leech. A parasite may lack or have lost certain organs, and therefore must live by using those of its host: mistletoe lacks roots, and fungi have no chlorophyll with which to manufacture sugars, so both suck nourishment from other, differently endowed plants. Such parasitism achieves its highest – or lowest, depending on your point of view – expression in the tropical, rose-like *Rafflesiales*, which, although a flowering plant, has no photosynthesizing equipment at all, and exists on the stalks of its host only as a flower. Anthony Huxley, in *Plant and Planet*, refers to fungi as "the misfits in the order of things," and although many of us don't seem to mind misfits much – we venerate truffles, which are the fruiting bodies of fungi – we don't really like to think about how they make their living.

I became aware of parasitism as a survival strategy by watching birds. The European cuckoo and the Brown-headed cowbird, for example, practise what is called brood parasitism; they make no nests of their own, but lay their eggs in other birds' nests, most often those of yellowhammers and yellow warblers (in the case of cowbirds), and let the host birds raise their young for them, often at the expense of the hosts' own offspring. (The New World cuckoo is less parasitic than its Old World counterpart, making its own nest and using it, mostly, and when it does lay its eggs in another bird's nest, the other bird is usually another cuckoo.) In Europe, the cuckoo is not despised: Shakespeare knew of its parasitic propensities, for he had the Fool warn King Lear that "the Hedge-sparrow fed the cuckoo so long that it's had it head bit off by it young," which is a lesson Lear, like the parents of my young friend, failed to heed. But the sight of a dozen cowbirds stuffing their beaks at my feeder while their surrogate child-care providers fly about frantically trying to feed their nestlings, fills me with righteous indignation. I have stood for hours on the deck of an icebreaker in the High Arctic watching Parasitic jaegers steal fish after fish from the beaks of industrious kittiwakes, and although I admired the gymnastic gall of the jaegers, I was pulling for the kittiwakes. But parasitism accords well with long species life: the old name for Brown-headed cowbird was Brown-headed bison bird.

We are appalled by parasites and applaud predators, but there is a thin ecological line between the two. A predator kills its prey and then eats it; with a parasite the order is simply reversed. This seems to be a morally ambiguous distinction. By another definition, it is predation when the "strong" kill the "weak," and parasitism when the "weak" kill the "strong." The weak preying upon the strong seems morally wrong to us, unnatural, un-darwinian even (the unfit triumphing over the fit), and it may well be this sense of injustice in nature that underlies our moral outrage at the presence of parasites among us.

For it is true that, as Stephen Jay Gould has observed, "nothing evokes greater disgust in most of us than the slow destruction of a host by an internal parasite," and nowhere is that disgust more heightened than when the host is us. Humans are more often the dinner than the diner. "Man," Darwin wrote in *The Descent of Man* in 1871, "is infested with internal and plagued by external parasites." The earliest medical text dealing with tropical diseases, by the Portuguese physician Aleixo de Abreu (1568–1630), dwells mightily on the internal bites that our flesh is subjected to. De Abreu had spent time in Angola and Brazil, where he contracted seven tropical diseases himself, and a large part of his treatise, published in 1623, is a description of the parasites within his own body that he had brought back to Lisbon and were still eating away at him. He described the *gusano Trichuris trichiura*, for example, a worm found in the rectums of victims of yellow fever; he also mentioned the *Tunga penetrans*, a flea that ate its way into his foot, entering under the toenail; and a Guinea worm, *Dracunculus medinensis*, which, he said, grows to the thickness of a violin string and "is extremely noisome to extract."

Darwin could not have made his observation about our internal parasites without the groundbreaking work of German zoologist Rudolf Leuckart, who, in 1862, published a series of studies on the tapeworm and the fluke. (His book on the subject, *The Parasites of Man*, was not published until

1879, and not translated into English until 1886, so Darwin must have read the earlier papers.) Leuckart was the first to trace the causes of a large number of diseases, from dysentery to trichinosis, to the vast array of parasitical organisms that swim in our blood and burrow into our flesh.

The English language owes a few words to parasitism: we call hangers-on "leeches" and "sponges," both of which are parasites in the natural world. They are both opprobriums. But a "fluke," in the sense of an unanticipated stroke of good fortune – the term originally meant sinking an uncalled shot in billiards – can be a positive thing. Most of the goals I score in pick-up hockey are flukes, and I am very fond of them. (In Vermont, I once saw a truck belonging to the Fluke Courier Company with the logo: "If it arrives on time, it's a Fluke!") In the human body, however, if it lodges in your duodenum, bores its way through your intestinal wall, tunnels into your liver, gets into your bloodstream, migrates to your brain, and kills you, it's a fluke. If a lot of flukes join together to form a long chain, if that chain then inhabits your lower intestine and grows to a length of thirty-two feet and produces larvae that migrate through your liver into your brain and kill you, it's a tapeworm.

Even parasites that technically have nothing to do with us we view with a certain frisson of distaste. Take the case of the Ichneumonoidea, a group of tiny wasps comprising some fifteen thousand species. The females have extraordinarily long ovipositors, which they use to inject their eggs into the larvae of other insects, usually caterpillars but also aphids and spiders. When the eggs hatch, the ichneumon wasp larvae begin to consume their host from the inside, being careful not to eat any of its vital organs, which would kill the host before the young wasps were done with it. They start with the fatty tissues, proceed to the muscles and digestive organs, and only when they are on the cusp of maturity do they consume the heart and the nervous system before piercing the inert body wall and flying away. The males emerge first and hang around to impregnate the females as they, in turn, wriggle out of their defunct host. Gould notes that the parasitical habits of the ichneumonids are not merely distasteful to us, but presented to nineteenth-century moralists "the greatest challenge to their concept of a benevolent deity," since they found it difficult to continue putting their faith in a God that had deliberately created such a gruesome, self-serving species of being, not once, but fifteen thousand times.

We are less inclined to moral censure, however, when it comes to forms of parasitism that benefit us. In this instance, parasitic wasps provide us with a precise year for the shift in our attitude toward parasitism: 1799. That year American entomologist William D. Peck discovered a wasp that injected its eggs into those of a slug worm. When it did so, the slug-worm eggs did not hatch. Peck erroneously called the wasp a slug-fly (wasps have four wings, flies have two), but he realized he had discovered an easy way of controlling slug worms, the major pest of pear and cherry trees: fruit growers had only to ensure that there was a good supply of slug-flies in their orchards. The benefit of parasitic wasps to agriculture launched the field of parasitology, and the hitherto distinct disciplines of biology and economics were yoked forever: five years after his discovery, Peck became Harvard University's first professor of natural history.

The slug-fly is known today as a species of *Trichogramma* wasp. Like most parasitic wasps, it is so minuscule it is almost invisible to the naked eye, which may be another reason for our tolerance of it. The insect has been intensely studied. Some female *Trichogramma*s will happily deposit their eggs in a variety of other eggs, including those of beetles and flies as well as caterpillars. In some experiments, they were even found trying to jab their ovipositors into egg-shaped grains of sand and globules of mercury. Such promiscuity makes it risky as an introduced parasite, since it could kill off other beneficial insects. Some of the ichneumonids, for example, are "hyper-parasites," which means they parasitize the eggs of other parasitic wasps. Other species are more selective and therefore more useful: *Trichogramma semblidis*, for example, specializes in the alderfly, whose carnivorous larvae bore into freshwater fish; *Ooencyrtus submetallicus* and *Asolcus basalis* both prefer the eggs of the green vegetable bug, and the former has been introduced in Australia to control that pest; the species *Apanteles congregatus*, very common in North America, attacks only tomato and tobacco hornworms. We like these beneficial little wasps; we study them, nurture them, ship them to various trouble spots around the world. We don't even call them parasites; we call them "biological controls."

New ones keep turning up. In the state of Morelos, Mexico, for instance, a new species of Strepsiptera – a small order of endoparasitic insect – has been found to do major damage to the corn leafhopper, *Dalbulus maidis*, a pest that causes such horrific-sounding diseases as corn stunt spiroplasma and maize bushy stunt phytoplasma, not to mention the dreaded maize rayado fino marafivirus. Although the leafhopper is already parasitized in Nicaragua and other parts of Mexico by a fly and two wasps, it is hoped that the new species will provide an effective control for corn growing in the higher altitudes of Mexico's interior.

I like this term "biological control" for the ring of optimistic confidence it lends to what is essentially a chaotic field of wild and exciting speculation. How do you "control" a parasite that will lay its eggs on anything that doesn't move and eat anything that does? *Trichogramma* species have been the objects of intense investigation since 1799, and they are still bafflingly unpredictable. *T. minutum*, for example, is well known as an egg parasite of the spruce budworm, *Choristoneura fumiferana*, and a study is underway to see how it performs on the closely related spotted fireworm, *C. parallela*, a pest of New Jersey cranberry bushes. Fireworm larvae "balloon" onto cranberry leaves and form massive leaf webs within which the worms eat away at leaves, fruit, and bog-owners' profits. Inconveniently, the female fireworm moth does not deposit her eggs in uniform clusters directly on the cranberry plant, which would make them easier to parasitize, but in masses ranging from three to three hundred eggs on twenty different species of plants that grow near the cranberry bog, most commonly red maple, several types of grass, leatherleaf, sweet pepperbush, and briar. So far the study has determined that, for some reason, the parasitic wasps show a marked preference for fireworm eggs laid in small clusters on sweet pepperbush and St. John's wort, and turn up their noses, if wasps have noses, at eggs laid on most other surfaces. Why this should be so remains a complete mystery. What a cranberry farmer can do with the information, other than somehow see to it that his cranberry bog contains loads of sweet pepperbush and St. John's wort, is equally perplexing. The interim report in the *Annals of the Entomological Society of America* concludes with the time-honoured scientific observation: "Further research is necessary."

Far from us using parasites to control our pests, there is some evidence that parasites may be the driving force behind evolution; in other words, that they may be controlling us. Carl Zimmer, in his book *Parasite Rex*, suggests that when parasites attack a species in sufficient numbers, they "push their hosts to becoming more diverse." In other words, increases in host-specific parasites act like climate changes or habitat loss; they are the change in an environment that forces the host species to modify its way of life enough, over time, to "become genetically distinct from the rest of their species." For example, human beings may be naked apes because hairlessness discourages body lice. And we may have moved into temperate zones to get away from tropical parasites. We owe a lot more to parasites than a few unblemished cranberries.

Further obscuring the traditional picture is the seemingly impenetrable mixing of genetic material since the Big Bang, which has made it virtually impossible to say which came first, eukaryotic organisms from which mammals eventually evolved, or single-celled parasitic prokaryotes. Heavy betting among geneticists these days is on the eukaryotes, which means that eukaryotes may have given rise to bacteria – which are, of course, parasites – before turning into us. This would mean that we are not derived from slime, as was previously thought: we may all have evolved from parasites. As Zimmer puts it, "parasitic animals such as blood flukes and wasps are practically our kissing cousins."

This softening of our attitude toward parasitism would be good news for my young friend if it weren't for one thing. In all of nature, there is no example that I know of in which a parasite preys on its own species. Even hyperparasites, which deposit their eggs in the eggs of other parasites, refrain from parasitizing their own eggs. To do so would be counter-evolutionary. You don't oviposit in your own offspring. That, in its most fundamental form, must be at the core of our loathing for human smell-feasts, toadies, leeches, and other mouth-friends: they parasitize their own kind, thus contributing, potentially, to our own eventual extinction.

So, the short answer to my young, party-going friend's question is this: There is nothing intrinsically wrong with being a parasite – provided you're the parasite, and I'm not the host.

THE ATCHAFALAYA SYNDROME

Τ he natural world . . . is every-
where disappearing before our eyes," writes E.O. Wilson in
The Future of Life. Since Neolithic times, he says, when the
first disgruntled nomad decided to clear a patch of ground
and plant a few beans, nature has been "cut to pieces, mowed
down, plowed under, gobbled up, replaced by human arte-
facts." Unlike any other species on Earth, since the end of the
last Ice Age we have been waging a ten-thousand-year war
against our own environment. And we have won, for what it's
worth. "We have killed off nature."

Wilson goes on to give another example of how we continue to behave like a grasslands species: we feel compelled to
cut down trees. A hundred and fifty years ago, Anna Brownell

Jameson, seeing piles of burnt-out stumps surrounding every homestead in Upper Canada, noted in her diary that "a Canadian hates a tree," but it isn't just Canadians, and it goes back further than the nineteenth century. Neolithic peoples, Wilson writes, "instinctively wanted the ancestral habitat, so they proceeded to create savannahs crafted to human needs." When we cut down trees, we are recreating an ersatz savannah, a landscape that pleased our eye and eased our anxiety. Europeans did the same thing when they arrived in North America and South Africa; they cut down trees and created mini-savannahs that they called farms and ranches.

I admit to having felt the compulsion to cut down trees for no good reason, much as Thoreau felt the urge to sink his teeth into a woodchuck. This spring, while I was reading Wilson's book in the mornings, I was out in our woodlot cutting next winter's firewood in the afternoons. I wasn't exactly cutting the place to pieces, but I was making a rather large and messy footprint on it, crafting it to human needs, because in dense woods, in order to cut down one big tree, you sometimes have to cut down a lot of little trees that get in the way. I note with regret the wording of that last sentence. As if in response to it, nature sometimes steps in and taps you on the shoulder and lets you know that news of its demise is still slightly premature.

That happened to me one afternoon when I cut down a tree that had been arched like a bow by the weight of ice during the ice storm of 1998, and had remained bent but still growing. Like Robert Frost's birches, once bowed by ice they remain bowed for years, looking "like girls on hands and knees that throw their hair / Before them over their heads to dry in the sun." This particular tree was thigh thick at chest height, rose about twenty feet above the ground, and then curved down so that its hair brushed the earth at a point some fifty feet away from its base. New branches had grown straight up along the arch, so that it looked less like a girl doing a cat stretch and more like a camouflaged Dimetrodon. When I cut it, though, at the root end, it went down like a cartoon drunk, with its feet and nose on the ground and its back arching sideways through the air. Unthinking, I stepped over the trunk, and the tree suddenly kicked up its butt end and caught me about mid-thigh (luckily) and lifted me high into the air. I felt as though I were astride a living, powerful being that was not very happy with what I had done to it. In my excitement, my hand gripped the chainsaw trigger, and the saw revved so rapidly that centrifugal force caused it to gyrate like the head of a large snake, trying to get at my neck. The tree rolled completely over and threw me, chainsaw still roaring, a dozen feet to one side. I flew through the air, somehow missing several other trees and large rocks, and landed on one foot, in one piece, with one hand still gripping the chainsaw. When everything was still and I had stopped shaking, I sat down and took new stock of my relationship to nature.

As Wilson points out, we have been battling nature since we climbed out of the trees and decided we liked grassland better. But what is it in nature that draws our ire? Perhaps it's the unpredictability that seems to us to be rampant in the natural world. We are a domesticated, order-loving species; we hate anything unforeseeable. Random acts of violence – drive-by shootings, people going postal in fast-food outlets, suicide bombers – are particularly devilish to us. They are gratuitous assaults against which we know we have no defence. Randomness is unintelligible to us, messy and counterintuitive. Emotions are unpredictable. We respond to emotional acts of violence – planes flown into the World Trade Center, for instance – with intellectualized acts of violence – the War on Terror – because our survival strategy as a species is to replace chaos with order.

The natural world does not appear to us to be an orderly place. Random acts of violence happen in it all the time, or at least so it seems to us. My tussle with the tree was an encounter with gratuitousness. That tree would have killed me as indifferently as a cow stepping on a newborn kitten. Our recent ancestors, I thought, as I sat on my fallen tree and contemplated the delicate moss patterns that stippled its trunk, must have experienced the sheer impartial unexpectedness of nature far more frequently than we do now. Subtract the chainsaw, and what had just happened to me must have been a fairly regular occurrence among men and women working in the bush in Anna Jameson's day. Trees falling in the wrong direction, "widow-makers" getting hung up in neighbouring branches, "barber chairs" splitting at the base and sending viciously pointed shards into exhausted sawyers' guts, boats hitting hidden rocks in rapids, barns and hay mows bursting spontaneously into flame. Not exactly freak accidents, but chilling warnings that human inattention and miscalculation – or the assumption that we know better than nature what is best for ourselves and it, that our needs are more important than, say, a forest's – are deadly affairs. Nature is occasionally forgiving, but we never know when it's going to stop overlooking our mistakes.

Every now and then the forest must have given them a little nudge, like an elephant leaning against its handler in a stall. The first would have been gentle, the next somewhat less so. The tree would fall a little closer to camp, or the blasted rock would hit a wagon on which no one happened to be sitting. Eventually someone would be hurt, and people would go around shaking their heads, or their fists, at the indifference and cruelty of nature. (A few would have said it was their own damned fault, but not many.) Not even Darwin was exempt from this view of nature as a quirky and malevolent force, a Dragon awaiting its St. George. For him, natural selection was pretty much a random process, depending as it often did on such unscheduleable natural agents as weather, floods, mountain and island formations, migration patterns, wind, birds, fur, driftwood, even icebergs. Anything that relies on an iceberg for its survival is shooting craps with Death. Nature, in Darwin's view, was vicious: in a letter to his friend (and eventual pallbearer), the botanist Joseph Dalton Hooker, he referred to "the clumsy, wasteful, blundering, slow, and horribly cruel works of Nature." The undeserved cruelty that seemed to be law in the natural world described by Darwin was the chief objection to his hypotheses put forward by his detractors during the Victorian era.

"I do not see," wrote John Burroughs, that gentlest of nature writers, in 1920, "that Nature is any more solicitous about the well-being of man than she is, say, about the well-being of trees." That's a humbling thought, and it was precisely mine as I went hurtling through space, chainsaw buzzing in my right ear. Nature doesn't care which one of us survives. Burroughs's book was called *Accepting the Universe*. In it, he asserted his belief in the essential goodness of nature. "It were a pity to go through life with a suspicion in one's mind that it might have been a better universe, and that some wrong has been done us because we have no freedom of choice in the matter." But is that not exactly what we do — imagine that there might be a better universe, that some wrong has been done us — when we impose what we think of as order upon what we think of as randomness in nature? And don't we almost always get it wrong?

On July 4, 2000, for example, Canadian biathlete Mary Beth Miller, while training outside Quebec City for the 2002 Winter Olympics, was attacked and killed by a black bear — she was found with bite marks on the back of her neck, there were black-bear tracks nearby, and a black bear had chased a group of cyclists in the same area earlier that week. Wildlife officers went out with their guns and killed a bear they thought was the villain — in other words, they responded to a random killing with a deliberate killing. A bear is terrorizing the neighbourhood: kill a bear.

DNA studies performed on the bear's claws and teeth, however, showed that the wildlife officers had shot the wrong bear. That in itself was remarkable; given the large territory black bears require for their home ranges, it seems unlikely that two would be frequenting the same area in midsummer.

But what was even more remarkable to me was the fact that the story made the six o'clock news. What was so newsworthy about wildlife officials killing an innocent bear? Not mass sympathy for the bear, certainly. Perhaps because it meant that the real killer was still out there, waiting for a chance to kill again. But the back-story was also interesting: we had responded to one random killing with a second random killing. We hadn't eliminated randomness, we'd added to it. That's what made us uneasy.

(It's worth noting at this point that, in a sense, all killing is random, depending on which end of the weapon you happen to be looking down. Astute readers will already have noted that my tree did not pick me to fling across the woodlot; I picked it. I could have cut down another tree, or none at all, but I chose that one according to a logic from which the tree, if it could think, would no doubt have tried to dissuade me. "If trees had powers of thought," wrote Burroughs, "what a struggling, agitated, unstable world they would think themselves born into!" As if insects, blights, fungi, and ice storms were not enough, my tree might have thought, now here's this idiot with a chainsaw! To the tree, I represented the randomness of nature. Similarly, the bear killed in Quebec was a victim of nature's randomness: like my tree, it hadn't done anything but mind its own business.)

The danger of answering chaos with what we perceive as order is the theme of John McPhee's book *The Control of Nature* (although his title seems to oppose the sentiment of Burroughs's *Accepting the Universe*, McPhee and Burroughs are actually on the same side). Thinking we can control nature used to be called hubris. McPhee quotes a former general counsel for the National Wildlife Federation, who calls it "arrogance." He also quotes an engineer with the United States Army Corps of Engineers — an outfit whose very raison d'être is to bend nature to the will of the United States government — to the effect that "whenever you try to control nature, you've got one strike against you." McPhee implies that it's more like a full count in the bottom of the ninth with two out and none on.

Still, that hasn't stopped the Army Corps of Engineers from trying, no sir. The U.S. Army Corps of Engineers was not formed to accept the universe as it finds it. In 1963, the Corps began trying to stop the Mississippi River from shifting out of its own channel into that of the Atchafalaya River just before it hit the Louisiana delta, as it is wont to do every two hundred years or so. For the past hundred millennia, the Mississippi has jumped from one path to another as it made its way to the Atlantic Ocean. Forty years ago it seemed on the brink of shifting again, and thereby debouching into the Gulf of Mexico a few hundred kilometres west of its present mouth. This would leave the city of New Orleans, a major seaport, an unacceptable distance from the sea. "For the Mississippi to make such a change," writes McPhee, "was completely natural, but in the interval since the last shift, Europeans had settled beside the river, a nation had developed, and the nation could not afford nature." The American Ruhr could not be allowed to become a tidal creek. "Nature had become an enemy of the state."

Well, as New Orleans residents have known for decades,

nature has always been an enemy of the state, but every now and then it slips its surly bonds and runs roughshod over shipyards and shopping plazas. The flooding of 80 per cent of New Orleans in the aftermath of Hurricane Katrina in late-August 2005 was only the latest in a long series of natural disasters that have affected that city. One of the worst was in 1927, when the levee system that held back the waters of the Mississippi River was broached in 120 places, and more than 165 million acres of Louisiana were flooded, leaving 600,000 people homeless. Huge deluges have occurred in the city on average every four years since.

In 1963, after the century's eighteenth major flood, the technicians were called in. The idea was to divert 30 per cent of the Mississippi's flow into a third system, the Old River, in the hope that doing so would take its mind off hopping into the Atchafalaya. McPhee put a name to this hubric behaviour. "Atchafalaya," he writes: "the word will now come to mind more or less in echo of a struggle against natural forces – heroic or venal, rash or well-advised – when human beings conscript themselves to fight against the earth, to take what is not given, to rout the destroying enemy, to surround the base of Olympia demanding and expecting the surrender of the gods." Our persistence in trying to force nature to conform to our own concept of order may be termed the Atchafalaya Syndrome.

Since McPhee's book was published, the Atchafalaya Syndrome has made a mess of the Atchafalaya River. The Corps of Engineers decided that the best way to stop all of the Mississippi River from changing course was to allow some of it to do so at places determined by the Corps of Engineers. As a result, 30 per cent of all the water flowing down the Mississippi River now flows into the Old River basin instead of some of it flowing into the Atchafalaya, which means that the Atchafalaya is now a vast, stagnant system of interconnecting swamps and marshes that somehow still has to support half of North America's migratory wildfowl, including 26,000 nesting pairs of herons, egrets, and ibises, and a fishery that expects 1,000 pounds of fish per acre. Environmentalists think 30 per cent is not enough. Water levels in the basin are too low, and the entire basin is silting up, which not only threatens the fishery and wildlife but, ironically, renders the catchment basin unlikely to be able to handle a major swelling of the Mississippi if and when a natural flooding event (like Hurricane Katrina, say) occurs.

Bill McKibben also believed that our war against nature has resulted in nature's near annihilation. In his apocalyptic book *The End of Nature*, he lamented that he could no longer admire a sunset without thinking that the rosy hue was probably caused by air pollution at the horizon. This is not new: in the nineteenth century, England burned so much coal that the French used to travel to London to admire the gorgeous sunsets over the Thames. But it is now virtually a global phenomenon. "We have killed off nature," McKibben wrote in 1989, anticipating Wilson's phrase by a dozen years, "that world entirely independent of us which was here before we arrived and which encircled and supported our human society. . . . Each cubic yard of air, each square foot of soil, is stamped indelibly with our crude imprint, our X."

By nature, he meant wilderness. The U.S. Army Corps of Engineers might like to think that wilderness has been defeated, but as McKibben points out, the new nature we have created is just as random as the old one was. "Simply because it bears our mark doesn't mean we can control it," he writes. "The salient characteristic of this new nature is its unpredictability." That has certainly proved to be the case with global warming, the salient features of which so far have been completely unforecastable weather, tropical storms turning unexpectedly into hurricanes, unprecedented forest fires, unaccountable droughts, unusual floods, and sudden rises or drops in temperature. The results, as we have recently seen in Louisiana and elsewhere, have been catastrophic. Raise the ocean-surface temperature by a degree or two and, to borrow Milton's description of chaos, all hell breaks loose. The old nature, the one we killed off because we didn't trust its fickleness, begins to look as dependable as clockwork by comparison.

Wilson agrees that nature is on the ropes, but as the title of his book suggests, he detects life in the old trooper yet. His has long been one of the most persistent voices raised against the wanton destruction of wilderness forests, and his are sobering words to ponder as one sits on a tree one has just cut down, surrounded by a dozen saplings one has just murdered for no better reason than that they were in the way, having just been gently told by the forest that one is on the brink of pushing it too far toward savannah. More than half of the Earth's trees are gone, Wilson tells us, and the remaining stands of what can still be called wilderness canopy are rapidly dwindling. The vast rain forests of Amazonia are 14 per cent gone and are disappearing at a rate of about 5 per cent a year (only 3 per cent of them are protected from logging). The great needleleaf forests of northern Canada and Russia, the temperate, Douglas-fir rain forest of the Pacific Northwest, the great forests of central Africa and New Guinea, are all threatened, all subject to the overriding arrogance that impels our species to modify whatever it sees to its purpose. Where we saw forest, we wanted prairies. Prairies are predictable, forests are random.

But Wilson and McPhee and even McKibben also live in hope, otherwise they wouldn't write books. In *Hope, Human and Wild*, McKibben points to several spots of sanity on the planet where attempts are being made to step more lightly on the Earth, to make a smaller X, where the Atchafalaya Syndrome seems to have been curbed. He cites Curitiba, Brazil, for example, which has banished internal combustion engines from its downtown streets, and the Indian province of Kerala, where overpopulation is being addressed by education. Even McKibben's home in the Adirondacks has seen the return of red wolves, new stands of pitch pine and post oak replacing pastures, and core samples from local bogs bringing up still-viable pollen from pre-European days. "If we can't prevent the environmental damage already underway," he writes, "we can – if we act boldly – limit it." To do that, we need to be motivated by more than fear. "To spur us on we need hope as well – we need a vision of recovery, of renewal, of resurgence."

Such a vision is what Wilson offers us. Science, he tells us, has made tremendous advances toward understanding that nature is not, after all, unpredictable and random. Fresh dispatches are coming in all the time. "As a biologist with a modern scientific library," he writes, "I know more than Darwin knew."

What those dispatches are telling us is that there is an underlying intelligence in nature. Chaos theory has taught us that even chaos has a kind of fierce logic to it. An orange is not clockwork, perhaps, but there is a certain determinism to its structure and shape. The British scientist Richard Dawkins has shown that evolutionary pathways — the patterns of spiders' webs, for example, or the internal structures of seashells — are predictable by computer. And the destruction that has led to our present environmental impoverishment is now well understood. According to Wilson, "we have a grip on its dimensions and magnitude, and a workable strategy has begun to take shape." It isn't too late. "Earth is still productive enough and human ingenuity creative enough" for the world to be saved, he writes, sounding a lot like John Burroughs. Except that Burroughs believed that what was required was a new recognition of our humanity, our innate goodwill. Wilson recognizes that it will also take a lot of money. If "one-thousandth of the current annual world domestic product," he calculates, "or 30 billion dollars out of approximately 30 trillion dollars" were to be dedicated to the protection and management of the world's remaining natural reserves, the great forests could be salvaged. Life would have a future. This could be accomplished, he calculates, "by a one-cent-per-cup tax on coffee." We could put our faith in Starbucks.

Meanwhile, events of what the American press has called "biblical proportions" – the tsunami that struck Asian coastlines in 2004, the devastation caused eight months later by Hurricane Katrina in New Orleans – serve to remind us that nature can shake off our controls pretty much at will, and does so regularly, like an Indian elephant deciding it no longer wants to lift logs but would rather chase villagers. Who's to stop it? Not the U.S. Army Corps of Engineers.

For my part, I've given up clear-cutting. I still cut the standing dead and the hopelessly bowed – heating with wood is a relatively small X – but now I remove the saplings with a spade and transplant them in places where I no longer want meadow. I even spare the prickly ash and poison oak. I have decreed our woodlot an Atchafalaya-free zone. We must work, as did Burroughs, to preserve the ultimate goodness of the universe, of ourselves, by ensuring that while the rule of might may have prevailed in the old world, it will be replaced in modern times by a newer rule, one which, according to Burroughs, "has come into the world and which is just as truly a biological law in its application to man as was the old law of might. I refer to the law of man's moral nature, the source of right, justice and mercy."

WHO OWNS A MOUSE?

One day in mid-winter I trudged out through the snow to check some tracks in our yard. Two deer had come out of the woods and crossed over to an apple tree in the far corner. It had been a poor year for apples and we'd left quite a few on the branches, small, deformed, scabby things, not even good for cider, and the deer were coming late in the evenings to get them. I could see where they'd raised themselves on their hind legs and struck at the branches with their forefeet, knocking the frozen fruit loose and then eating them off the ground. Picturing this, I realized that while they foraged for the apples they were exposed to all kinds of anxieties in our yard, from passing cars and neighbours' barking dogs to more natural predators, such as coyotes. I began to pick the apples from the branches and toss them into the woods, so that the deer could eat them without having to leave the shelter of the trees.

And then I stopped. If the deer didn't leave the woods, I thought, I wouldn't be able to see them when they came for the apples. I wanted to see them on their hind legs, hoofing the tree, the male rattling the branches with his antlers. I wanted to watch as they bent their elegant necks to pluck the fallen apples from the snow. Just one night more, I thought. After I'd seen them I'd come out and toss the rest of the apples into the woods.

I'm not proud of that thought. By leaving the apples on the tree, I was luring the deer out of the safety of their natural habitat into my orchard, solely so that I could stand in the comfort of my kitchen and gaze at them through the window. What arrogance! I was behaving as though I owned the animals.

Wanting to see wildlife isn't in itself an unnatural or shameful desire. For many of us the chance to observe wild animals in their natural habitat is one of the joys of hiking or canoeing. It's more than just wanting to see something beautiful. Novelist and art historian John Berger has thought deeply about the importance of looking, and according to him, looking at animals (and being looked at by them in return) is one of the most visceral experiences we can have. The first element of language, he believes, was metaphor, and the first metaphors were of animals. People were as brave as lions or as shy as deer, as swift as gazelles or as cunning as foxes. Animals were self-referential: they were not only like us, they were us. He quotes Claude Lévi-Strauss quoting a Hawaiian elder: "We know what animals need, because once our men were married to them and they acquired this knowledge from their animal wives." Similar foundation stories exist among the Haida-Gwai and Coast Salish people of the Pacific Northwest. They also crop up in Greek mythology, in stories in which gods change themselves into animals and consort with humans, or in which humans are transformed into trees and flowers. In modern times, says Berger, animals have become marginalized; machinery has replaced oxen on farms and horses on the road; meat now comes in plastic wrap, preseasoned on supermarket shelves. Animals confined in zoos attest to our curiosity, not our sense of oneness: "The zoo cannot but disappoint."

Going into the wilderness to be among wild things reminds us of our own former wildness. It puts us back where we like to think we belong. It sharpens our instincts, hones our survival sense. But Berger is right: wild animals have been marginalized from our lives. They have disappeared as we have become more civilized. Now when we see a wild animal we are like a domestic cat watching a squirrel through a kitchen window. Our ears perk up, our tail bones flicker, a kind of deep, predatorial kecking sound starts in the backs of our throats, and for a brief, glorious, uncontestable second we, too, are wild and free, crouched in our rightful place in nature.

Like the cat in Stanley Kunitz's poem "Raccoon Journal," however, we have become "domesticated out of nature." His cat

. . . stretches by the stove
and puts on a show of bristling.
She does that even when mice
Go racing round the kitchen.
We seem to be two of a kind . . .

Wanting to control the animals we see — as I did with the deer in my orchard — is the beginning of domestication. It is a desire to own that animal. Leaving those apples on the tree was a step down the road that leads to the Harvard mouse.

The Harvard mouse has never seen a field or a forest. It doesn't know what shelter means. It hasn't been merely domesticated out of nature; it has been manufactured out of nature. It's a laboratory strain, probably a derivative of Black 6 — your average, run-of-the-mill lab mouse that has been used for years in medical experiments — that has been genetically altered to be highly susceptible to cancer. It's also known as the oncomouse.

To make a Harvard mouse, researchers inject a cancer plasmid with the oncogene, known as "myc," that is responsible for susceptibility to cancer (women whose mothers have had breast cancer know about this gene, because they have probably considered having it removed from their own daughters). The plasmid is injected into a female Black 6's fertilized egg while the egg is in its single-cell stage, then the egg-plus-plasmid is transplanted back into the female Black 6 and allowed to grow to parturition. The resultant newborn is called the "founding mouse": all its cells are susceptible to cancer. The founding oncomouse is then mated with a normal Black 6, and half of their offspring will contract cancer and grow tumours very rapidly when exposed to something carcinogenic. The other, normal, half is the control group. It's a useful animal. It saves researchers months of valuable time.

It's also a profitable piece of property. Harvard University patented the Harvard mouse in the United States in 1988, making it the first transgenetically engineered organism to be patented anywhere in the world. Harvard now owns the Harvard mouse in exactly the same way that Disney owns Mickey. Since 1988, any research lab wanting an oncomouse has had to buy it from Harvard, or, more accurately, from a breeding house licensed by Harvard, such as Dupont. Most labs happily pay up, because the Harvard mouse has become a kind of status symbol in the cancer-research field. A lab with a spanking new Harvard mouse can test a drug manufacturer's product more quickly than one with an old Black 6, and so Harvard-mouse-equipped labs attract millions of dollars in research funding and pharmaceutical contracts from drug companies trying to beat one another to the pharmacy shelves. Universities tend to approve of such boons to humankind, much preferring them to the other kind of boons, the kind that don't translate into huge grants and lucrative contracts. Advances in comparative literature, say, or fresh insights into medieval atonal music.

Or ethics. Questions about the ethics of genetic research are muted, because they are embarrassing and retard progress. The Harvard mouse raises all kinds of them. It is, in a sense, the founding mouse of all ethical qualms about the patenting of living organisms for private profit. The patent granted to Harvard University opened the gates to a flood of patents for such products as genetically modified canola, soybeans, corn, cotton, potatoes, tomatoes, rice, sugarbeets, and wheat. The list goes on and is growing annually. Since 1990, the Canadian Patent Office has received applications for patents on 250 genetically modified animals, plus another 350 for genetically modified plants. Genetically Modified Organisms are a growth industry, and the Harvard mouse is the granddaddy of all GMOs.

Harvard applied to the Canadian Patent Office for a licence to sell the Harvard mouse in Canada in 1985, and the patent pended until 1993, when the Canadians finally rejected the application. It was one of the most monumental ethical decisions ever to go unnoticed in Canada. The patent office said that it patented "inventions," defined by the Federal Patent Act of 1869 as "any new and useful art, process, machine, manufacture or composition of matter," and that the Harvard mouse was not an invention. Harvard's patent lawyers argued that the Harvard mouse was "a composition of matter," but the Patent Office wasn't impressed. Anything from a hay fork to a human can be defined as a composition of matter, but it was not the intention of the Patent Act to allow the patenting of human beings. The Commissioner of Patents decided that Harvard could patent the process of making a Harvard mouse – no other lab could go about making an oncomouse the same way that Harvard did – but it could not patent the resultant mouse itself. The mouse was a living organism, a product of nature, it grew from a cell that became an embryo that became a mouse in the usual way, and was not an invention of Harvard. Composition of matter might conceivably apply to single-celled organisms (like yeast or bacteria or the single cell that grew into a multicellular oncomouse), but it could not apply to anything more complex than that. Hurrah for the Patent Office, I say.

There is a wrinkle in this story, however. In August 2000, pressure from biotechnology firms such as Monsanto caused the Canadian government to appeal the Patent Office's decision, and the Court of Appeal overturned the 1993 ruling, allowing Harvard to patent the mouse as well as the process in Canada. Harvard's lawyer, David Morrow, called the decision "the triumph of logic and reason over visceral fear." (For "logic and reason" read Science, for "visceral fear" read Religion.) In a remarkably characteristic Canadian moment, the Canadian government then appealed *that* decision to the Supreme Court, and on December 5, 2002, the Supreme Court overturned the 2000 decision, and unpatented the mouse, making Canada the only First World nation that does not allow the patenting of animals. Huzzahs for the Supreme Court, too. Visceral fear usually wins out over logic and reason hands down, or what's a Heaven for?

Morrow had argued before the Supreme Court that the Patent Office erred on the side of arrogance in making a distinction between unicellular and multicellular organisms. "In the spectrum of life on Earth," he said, "there is no difference between a mouse and a bacterium. It's a continuum. Where do you draw the line? Do you patent the squid but not the shark? Or do you patent the shark but not the porpoise? There isn't any bright line."

The sublime speciousness of that argument apparently did not escape the judges. Biologically speaking, in the spectrum of life on Earth there is a very bright line between a bacterium and a mouse, and we should not be afraid to draw it. A bacterium is composed of a single cell – it is a prokaryote – and a mouse, like a squid or a shark, is a eukaryote, composed of a great number of cells. It may appear to be somewhat arbitrary, ethically speaking, to allow the patenting of the former and not of the latter, but at least it's a line and the reasoning is clear. Thus the rhetorical question, "Do you patent the squid and not the shark?" is easily addressed: You patent neither the squid nor the shark, both being eukaryotes. Ditto the porpoise. And ditto corn, canola, potatoes, and all the rest of them, all eukaryotes. (Although in a decision reached in 2004 the Canadian Supreme Court allowed the patenting of plants. So much for logic and reason.)

The mention of porpoises is strategically interesting, since it introduces the notion of animal intelligence. Porpoises are supposed, at least by lawyers, to be more intelligent than sharks, because they have little grey cells in their brains and a highly developed system of communication. Somewhere near the top of the Great Chain of Being, like a strand of genome code in the spectrum of life on Earth, there is a sequence that reads, in ascending order: shark, porpoise, dolphin, whale. Then there's a gap, followed by, before we get to humans, another strand that reads: apes, orangutans, rhesus monkeys, chimpanzees, bonobos. It's a long chain, but they're getting it straightened out for us. It's a continuum, you see, within which it is sheer cussedness to distinguish between a bacterium and a human being.

But there is an aspect of the oncomouse that makes cussedness the appropriate stance to take in this discussion. When making a Harvard mouse, the myc or oncogene that is injected into the cancer plasmid is not a mouse oncogene, it's a human oncogene. The thinking behind this is that a Harvard mouse with human oncogenes will respond to carcinogens in a more human-like fashion than one with mouse oncogenes. The tests will be more valid; that is, more likely to attract venture capital to further develop the drug. As one cancer researcher explained to me, human oncogenes and mouse oncogenes are 95 per cent similar, but that 5 per cent is critical. You might develop a drug that does not give an oncomouse with mouse oncogenes cancer, only to put that drug on the market and find it gives people with human oncogenes cancer. Better to find that out at the mouse stage.

But doesn't that raise questions about the true distinction between a mouse and a human being? To my mind, there is no bright line between mice and humans. This is especially true of an oncomouse, which has been spliced with human genetic material. Every cell in the adult mouse has a bit of *Homo sapiens* in it. When Harvard University patented its oncomouse, it was, in effect, patenting a piece of human tissue.

This may seem a rather academic point, but it has been given pith and moment by the recent announcement in the scientific journal *Nature* that researchers have succeeded in identifying the mouse genome sequence. This is big news. It came only one year after the publication of the human genome sequence. Now that geneticists know the mouse sequence, they can make comparisons between us and mice that will help to explain what each human gene does. We have 2.9 billion letters in our sequence; mice have 2.5 billion. But both mice and humans have thirty thousand genes, and only three hundred mouse genes diverge from those of humans. So, genetically speaking, mice differ from us by a mere 1 per cent. This small difference looks even smaller when you consider that one human being differs in genetic makeup from another human being by only 0.2 per cent.

The similarity between mice and humans means that mice can be fairly easily "humanized" by genetic modification. Now that their sequence and ours are known, mice can be made to grow any human gene, not just human onco-genes; it has been suggested that they be given human brain tissue, for example, or made to produce human sperm. Peter Bork, a molecular scientist in Heidelberg, Germany, has admitted that such modification "would take a hell of a lot of work," but he sounded as though he were rolling up his sleeves to get down to it. Before too long, it seems, the world's patent offices are going to be deluged with applications for patents on mice with miniature human body parts – little livers, tiny hearts, picayune pancreases – so that we can more accurately and swiftly test the efficacy of a whole new range of pharmaceutical products without using humans as guinea pigs. We can turn guinea pigs into humans.

Because why stop at mice? Since we're on a continuum, if there's no real difference between a mouse and, say, a chimpanzee, why split hairs? At last count, there were more than sixteen hundred chimpanzees living in American medical research facilities, most of them being used for HIV and hepatitis experiments. The chimps are injected with HIV-positive blood and monitored to see if they come down with full-blown AIDS. Very few of them do. In one study, of two hundred chimps infected with HIV, only two died of AIDS. In another well-documented case, a chimp named Pablo, kept at the New York University lab, was injected with ten thousand times the lethal dose of HIV and given four different test vaccines, as well as twenty-eight liver, two bone-marrow, and two lymph-node biopsies. The result: no sign of AIDS and only a mild case of hepatitis. He died, at the age of thirty in a primate retirement facility near Montreal, of infections caused by the 220 tranquilizer darts he'd been shot with over the years.

Obviously, chimps as chimps don't make good test subjects for AIDS research, which is why most labs in the world are looking for alternatives. But what if chimps were "humanized," genetically modified to grow human body tissue? If we can do it with mice, how long will it take researchers to think about doing it with chimps? Will there be, in some not-too-distant future, a laboratory strain of chimpanzees with human livers, human brains, human hearts? What would such an animal be like? More than one hundred years ago, T.H. Huxley wrote that "every principle gyrus and sulcus of a chimpanzee brain is clearly represented in that of a human," and he has been proven right. Although human brains are three times bigger than chimp brains, the frontal lobes in chimps and humans — the part used for complex mental activities such as language, artistic expression, and decision making — are almost identical in relative volume and cortical surface. In both human and chimp brains, each neuron in the cerebral cortex allows for a possible 10 billion billion synaptic connections. If you're going to argue that there is "no bright line" between a bacterium and a mouse, then you'd have to admit that the distinction between a chimp and a human is largely one of lifestyle.

Indeed, it was chimps' similarity to humans that made them attractive as test subjects for human drugs in the first place. That they have turned out to be less than ideal is puzzling to scientists; it must come down to only minor differences, but we don't know what they are. So how long can it be before someone decides that rather than try to overcome those differences for the purposes of medical research, perhaps it would be easier to eliminate them?

These may be unanswerable, even ridiculous, questions, but I believe they are worth asking. Barry Commoner, the maverick biologist who in the 1970s revealed the connection between chlorine and cancer and now heads New York University's Critical Genetics Project, is concerned at the degree to which "technology has gone beyond what science knows." Biotechnicians are churning out hundreds of genetically modified organisms — from herbicide-resistant soybeans to nutrient-enhanced tomatoes — without having the slightest idea what their effect on the human environment will be.

"No one," he told me, "is giving serious thought to the fact that they are transforming genetic material over which they have no control when they put it out in the field. That is what I worry about."

Well, the Canadian Patent Office has given it serious thought, and that's encouraging. But all the patent officers really did was turn the question back to the government, which handed it over to its Biotechnology Advisory Committee, a group of experts convened in 1999 to recommend government policy on issues involving genetic modification. Since all the members of the committee are scientists representing biotech labs or corporations, it may be only a matter of time before changes to the Canadian Patent Law will allow the patenting of a whole ark full of genetically altered animals as well as plants.

And while we're giving serious thought to the consequences of the genetic modification of animals, might we not wonder what kind of creatures are being contemplated in those 250 patent applications awaiting approval from the Patent Office? Might we not ask ourselves what provisions are being made to prevent such creatures, when they come to exist, from escaping into the wild? Could we, on some not-so-future hike into the wilderness to observe nature at close range, come upon some hepatitis-ridden rodent with a human brain? When I asked one cancer researcher what would happen to a Harvard mouse if it suddenly found itself in a sunny meadow instead of a laboratory cage, he answered, "It wouldn't survive." He did not say, "It can't happen." It can happen. It does happen. There is an island off the coast of

Liberia on which hundreds of "retired" research chimpanzees, all of them infected with strains of HIV and hepatitis, have been let loose to enjoy their declining years in the comfort of their own habitat. Are we worried yet?

Meanwhile, it strikes me as interesting that laboratory strains of mice have common names like Black 6 and Harvard mouse, but don't seem to have any scientific name. They are not *Peromyscus maniculatus* (deer mice), nor are they *Peromyscus leucopus* (white-footed mice). They're just mice. The onco-mouse, though it differs significantly from all other mice, is a taxon without a name. Linnaeus must be twitching in his grave. In order to clarify their position in the Great Chain of Being, I suggest oncomice be given a true taxonomic identity. I propose *Carcinomyscus huxleyii*. Let scientists think it is named for the great nineteenth-century biologist Thomas H. Huxley, champion of the lower orders. You and I will know that it is really named for T.H.'s grandson Aldous, the author of *Brave New World*.

MAKING A GARDEN

In Italy one May, my wife, Merilyn, and I rented a silver Peugeot convertible and, with the top down and our hair (well, her hair) flying in the Tuscan wind, we wove through the beautifully pastoral countryside, past rolling fields flanked by soaring cyprus trees, and through dozens of walled, hilltop villages with names I had formerly known only from literature and the labels on bottles of red wine: Montepulciano, Montalcino. The countryside was as green as chlorophyll could make it, all neatly partitioned into chequered squares; pasture, vineyard, orchard. It was a landscape virtually unchanged since the Middle Ages, in some cases since early Roman and even Etruscan times; a landscape nevertheless shaped by humans, and on a scale we are only beginning to see in North America. A pastoral landscape is a drastically altered one. Rarely did we see anything resembling a forest, and even when we did, what counted as a forest in Tuscany amounted to little more than a woodlot to our North American eyes. And a carefully managed woodlot at that. In Tuscany, indeed in much of Europe, what to the local inhabitants is wilderness to us seems more like a garden.

We drove into Florence. Actually, one doesn't drive into Florence; one approaches it on the bias, as a fox approaches a thicket, circling, sniffing for a break in the wall. The Porta Romana presented itself as a traffic circle complicated by mopeds and bicycles. Inside the wall, the city was a fascinating series of paradoxes. While Merilyn sauntered through the Boboli Gardens, I flannered my way through a pair of ordinary wrought-iron gates into the courtyard of the University of Florence. I had actually set out for the Uffizi, but such is the nature of flannering that I sidestepped into something else. Once within the university gates I passed a rank of parked mopeds, climbed three flights of worn stone stairs, and sat in a room dedicated to Galileo, who had lectured there in 1610. Sitting among a smattering of students, gazing at the vivid frieze of scenes from Galileo's life that surrounded the room at ceiling level, I found myself thinking about the momentous clash between science and religion that took place in the Europe of Galileo, and of which he became the focus. He lost his job at this university for defending Copernicus's conviction that the Earth moved and the stars did not; the Church preferred Aristotle's belief in an immutable universe, formed once and remaining thereafter forever unchanged. Since the sixteenth century, science and religion have continued to hold conflicting and seemingly irreconcilable views. For example, science tells us that humans evolved from wilderness dwellers to become gardeners; religion contends that we were created in a garden and evicted from it to become wilderness dwellers. Can anything, I wondered as I listened to a professor lecture in Italian, heal such a rift as opened up in this room?

To consider the question properly, I needed to start with some definitions. What is wilderness? And what is a garden?

One flight up from Galileo's lecture hall is the University of Florence's Museum of Natural History, which occupies the entire top floor of the building. It contains, among many other things, the famous "plague waxes" from the seventeenth century, four boxed dioramas – "The Pestilence," "The Triumph of Time," "The Corruption of the Flesh," and "The Gallic Disease" – by the Sicilian artist Giulio Gaetano Zumbo, each depicting miniature nightmarish scenes of the Black Death: bodies piled on bodies, lovers weeping over the emaciated corpses of their beloved, babies sucking the breasts of their dead mothers. Could anything in Aristotle, I wondered, prepare a world for this corruption in nature? Could anything in science? The museum also houses an astonishing collection of perfectly preserved eighteenth-century anatomical teaching figures – a room of legs in every state of dissection (the muscles, the circulatory system, the nervous system), others containing a wall of opened hands, a sea of flayed faces, an ocean of gaping thoraxes, each so lifelike that I didn't realize until I read the brochure that they, like the

Zumbo dioramas, were made of wax. They too were works of art, though they were created for science.

In other rooms, there were hundreds of stuffed animals from around the world, only slightly more recently added. Whoever put the collection together had an affinity for gazelles, but there was one astonishing room containing nothing but a stuffed rhinoceros. Another included a Tasmanian wolf. There were rooms of primates, which gave me the extremely creepy sensation of being observed by my unknown next-of-kin: not the chimpanzees and bonobos in which we recognize our former selves, but our much more distant, diminutive relatives: the DeBrazza monkey (*Cercopithicus neglectus*), a mantled colobus (*Colobus guereza*), and the tailless siamang (*Holobates syndactylus*). There were five enormous rooms of birds, one of which contained the head of a Dodo. It was in a case labelled *Etinte* (Extinct) with a dozen other birds, among them a great auk, a Black emu, and a passenger pigeon.

Walking among the glass cases, each containing pairs of dead, accusatory eyes, I couldn't help but think of the vast stretches of wilderness that had been emptied to supply them. The sheer numbers suggested it: could there be any left in their natural habitats? To the European consciousness, formed over centuries in a land where true wilderness no longer existed (Schiller noted in 1796 that our feeling for nature is like "the feeling of an invalid for health"), nature was elsewhere, out there, a vast, distant storehouse from which items of use or curiosity — a pair of Thompson's gazelles, a stand of Douglas firs — could be removed without any permanent damage being done to the whole, or none that could affect us. Wild nature was a museum; somewhat chaotic, perhaps, but full of interesting things. The animals and plants arrayed there had been placed on display by some Cosmic Curator for our amusement or enjoyment. The museum was a place we visited but did not live; it didn't have much to do with our ordinary lives. And the thing about a museum was, if you dismantled one or two displays, you still had a museum.

To some degree, unfettered nature is still regarded uneasily in Europe, like an intruder at a garden party. Think of those manicured woodlots with all the weeds removed. When landscape architect Antonio Perrazo was commissioned by an Italian electric company to design a garden that would illustrate the company's dedication to the environment, Perrazo proposed installing a patch of bare earth which nature could fill with whatever happened by – weeds, trees, grass, and wildflowers; eventually, he said, the patch would become a naturalized wood. The company rejected his proposal, claiming that it would "pollute the landscape." The irony was not lost on Perrazo, who thought the company was really balking at handing over a hefty fee when it was Nature that was doing the work. There were no objections to transplanting a thousand-year-old olive tree from Sicily to make "an instant garden," he said, but "who would pay an architect to make a wood?" It's probably true that a patch of ploughed land left to become naturalized would soon fill up with introduced species, and so be no less artificial than a planted garden, but at least that was a natural process, each plant growing from seed to seedling to maturity in its own time and manner.

But the European (and now North American) view of nature is of something foreign and dangerous. Nature is the meteorite in Peter Hoeg's novel *Smilla's Sense of Snow*, an alien rock that harbours a deadly virus. Nature is the cloned dinosaurs in *Jurassic Park*, which are all right as long as they remain on their remote island. Let raw nature have contact with civilization – let the bull into the china shop – and chaos ensues. (The virus gets out; the dinosaurs escape from the island.) But by the time the New World entered the European picture, nature was so reassuringly distant it was considered safe for plundering (for study, or for food, or for sport) without doing it lasting damage. John James Audubon, the nineteenth-century nature artist, killed dozens of guillemots and puffins in order to paint them "from the life." The wilderness couldn't be harmed; there was so much of it. If we created a vacuum, nature would fill it (as in Antonio Perrazo's brilliantly sarcastic proposal). To describe North America, even Thoreau used words like *infinite* and *inexhaustible*.

However, "there came a tipping point," as E.O. Wilson writes in *The Future of Life*. "By the time the American frontier closed, around 1890, wilderness had become a scarce resource at risk of being eliminated altogether." Nature, it seemed, was no longer closing in the gaps we were busily creating. Nature was in retreat, not just in Europe but in North America, and it wasn't only the animals that were disappearing, it was also the plants, of which there were far more species to plunder. The numbers are there if we need them: 19,078 plant species alone are now threatened with extinction. They are disappearing at the rate of two per hour.

The 1964 Wilderness Act defined wilderness as a place where "the Earth and its community of life are untrammeled by man and where man himself is a visitor who does not remain." Does any such place exist today? As early as 1864, George Perkins Marsh, in his book *Man and Nature* – sometimes called "the fountainhead of the environmental movement" – noted that "the destruction of the woods was man's first physical conquest, his first violation of the harmonies of inanimate nature," and that violation has been going on at the rate of about 2 per cent per year ever since. Not many untrammelled woods remain. Wilson guesses that the rain forests of the Amazon, central Congo, and New Guinea still qualify as wilderness, and the coniferous forests of northern Canada and Eurasia. He also includes "Earth's ancient deserts, polar regions and open seas." He does not, as some do, consider the concept of wilderness a thing of the past, that the museum is shutting down for good. Wild places still exist: "Walk from a pasture into a tropical rainforest, sail from a harbour marina to a coral reef, and you will see the difference," he maintains. "The glory of the primeval world is still there to protect and savour."

Well, Wilson is struggling against pessimism. He doesn't want us to give up. But natural habitats are not necessarily wilderness areas. And if you believe that global warming is a direct consequence of our activities since the Industrial Revolution, then there is not a square metre of the planet that is "untrammeled by man." The very existence of the northern boreal forest, breeding grounds for three hundred of North America's songbird species, is threatened by global warming. Nor are humans mere "visitors who do not remain" in the polar regions and open seas. I have been to the North Pole with a team of ocean scientists, and I have watched as they measured high levels of chemical and radioactive contaminants in the water, the air, and the biota, permanent markers of our civilization that have penetrated to the remotest regions.

But let us agree that there is hope and that Life has a future, that true wilderness areas do exist, even in North America. If they are to continue to do so, then they must be protected. Each one of those endangered plants must be saved. Not by legislation, which we now know doesn't work; they must be actively nurtured and, where necessary, repropagated. And that will make their wilderness habitat a garden.

The earliest gardens were walled. The word *garden* comes from the Old English word for *enclosure*, and is cognate with the word *yard*. The gardens we saw in Italy, notably the public Boboli Gardens in Florence, but also most of the smaller private gardens, were walled. The Bible is mute as to whether or not the Garden of Eden was walled, but it was planted – "The Lord God planted a garden eastward in Eden" – and later commenters certainly envisioned it as a walled enclosure. The fifth-century poet Avitus, for instance, who was also the Bishop of Vienna, wrote that

East of the Indies, where the world begins,
where earth and sky are said to meet together,

a grove stands, inaccessible to mortals,
on heights surrounded by eternal walls . . .

and his contemporary, Dracontius, a Carthaginian lawyer,
described Paradise as the "happiest garden in God's universe,"
surrounded by "a wall of densely interwoven branches." It was
a particularly unnatural place, a locus of "perpetual spring,"
where "bees need not manufacture waxen cells." (The conceit
is: if it's perpetually spring, bees don't need to store honey, so
they don't need to make hives, which is to me a slightly dis-
comforting concept. It reminds me of my first trip to the
Caribbean; driving through the countryside on Dominica I
saw lots of cattle but no barns. It took me a while to realize
that with a twelve-month growing season they didn't need to
store hay, and it was that more than anything that made me
feel I was in a foreign country.)

This business of the wall around Eden has significance in
North America. If Eden was planted, it was domesticated;
Adam and Eve were cast out of order into disorder, which
they then had to tame and cultivate in order to survive. When
European explorers came to the New World, they reported
that they had found "an Earthly Paradise," but they must have
known that what they had really found was wilderness. As
early as Columbus's second voyage, in 1494, one-third of his
crew fell sick with either syphilis, malaria, or bacillic dysentery. They also nearly starved to death. Kirkpatrick Sales, in
The Conquest of Paradise, finds this "almost unbelievable in
retrospect, if we consider the bountiful menu they had before

them: cassava bread, sweet potatoes, corn, peppers, peanuts, fish of all kinds, clams, conches, turtles, papayas, pineapples, plums, pears, and so on." Disease, famine, death: what kind of Paradise was that? It sounds more like the nightmarish scenes depicted in the Zumbo dioramas. "The bizarre refrain of hunger and malnutrition," continues Sale, "would be almost ceaseless during these early years of settlement, until the colonists finally determined which *European* crops would grow in the new lands." That is, until the settlers had succeeded in domesticating the wilderness, in turning it into a garden. As the sixteenth-century Spanish humanist Hernán Perez de Oliva observed, the New World was a place to be conquered in order "to give those strange lands the form of our own."

We now like to think of wilderness as a refuge, a solace, a place to go to replenish the soul, much as Europeans once thought of gardens: "Gardens delight, divert, support, and nourish," wrote the Latin poet Asmenius in 400 AD. That doesn't sound much different from Thoreau, himself no mean gardener, who wrote of wilderness as the ideal environment in which "to settle ourselves, and work and wedge our feet downward through the mud and slush of opinion and prejudice and tradition and delusion . . . through Paris and London, through New York and Boston, till we come to a hard bottom and rocks in place, which we call reality." How is it that by the mid-nineteenth century we had come to think of the two biblical opposites – garden and wilderness – in the same terms? Perhaps because, after 350 years, we had succeeded in turning wilderness into garden. And in the 150 years since Thoreau the process has continued: when wilderness has to be protected, as it now does, it becomes a garden without walls. It becomes domesticated.

But there are those intent on reversing the process, and they are not, as we might expect, wilderness dwellers; they are gardeners. In *Human Scale*, Kirkpatrick Sale has also written that "to fully and honestly come to know the earth, the crucial and perhaps only and all-encompassing task is to understand the place, the immediate, specific place, where we live." He was talking about gardening, since "where we live" is not, by definition, the wilderness, which is a place we visit and do not remain. No one knows the earth like a gardener. As Karen Landman, who teaches landscape architecture at the University of Guelph, points out, "people who garden know the soil, the weather, the climate, they know which species thrive naturally and which need nurturing." That would seem to fulfill Sale's requirements. Gardeners know precisely how many millimetres of rain and hours of sunlight fall on their gardens, they know about insects and fungi, animals and birds. They know what zone they live in. Eventually, says Landman, "they come to know themselves through their intimate connection with nature."

In her practice, Landman works at creating "natural" gardens, urging the use of local materials in making a garden. Local stones, local trees and shrubs, native plant species. She has become adept at ecological restoration, which is the attempt to return denaturalized land back to its original state before settlement by Europeans, or at least get it as close to that ideal state as possible. "You can never restore it fully to its natural state," she says. "It's far too late for that. But we can point it in that direction, and we can stop preventing it from getting there on its own." She enthuses about the work of Mathis Natvik, for example, of Orford Ridges Native Plants near Muirkirk, Ontario. Natvik's idea, called the Clear Creek Pits and Mounds Restoration Project, is to return a section of land that has been intensely farmed for two centuries back into a slice of Carolinian forest.

In our drive to turn forest into savannah, or wilderness into garden, we cut down trees and levelled the ground with earth-moving equipment. "When settlers came," Landman says, "they cut and burned the bush and ploughed the land, sometimes twenty or more times in the first year, to level it out for farming." The result was a landscape drastically altered, not merely visually but also viscerally. "The original forest floor was a random series of pits and mounds; when trees fell over, their root systems created large hillocks, and the valleys between them filled with water and became pits. Different species of plants grew on the mounds, which were well drained and received more sunlight, than in the pits, which were poorly drained. Squirrels would bury nuts in the mounds, and trees would grow from them, and the pits between the trees would grow ferns and creepers and other water-loving plants." Farming levelled the land and destroyed the species that had evolved along with it; Natvik used a bulldozer to return it to pits and mounds, and hired local schoolchildren to plant such Carolinian species as hickory, basswood, hackberry, red ash, and tulip trees as well as a variety of herbaceous perennials native to the region.

Great care must be taken in the selection of native plants, Landman warns. The goal is restoration, not recreation – she wants to make a place where native plants can begin to reclaim their birthright, not merely create something artificial that looks natural. The place will know the difference. For example, one of the species originally found in Carolinian forests is the American high-bush cranberry (*Viburnum trilobum*), a four-metre, leafy bush that produces bright red berries in the fall that remain on the shrub throughout the winter. It grew naturally here, but early European settlers brought in the Eurasian variety (*Viburnum opulus*), which spread wildly and has almost completely ousted the native bush. The two species are so similar that horticulturalists need a magnifying glass to tell them apart (the native variety has club-shaped glands at the base of the leaf, while the alien species has glands that are more concave). Birds that feed readily on the American high-bush cranberry, however, do not feed at all on the Eurasian variety. And insects, particularly the Viburnum leaf beetle, which can defoliate a cranberry bush overnight, show a preference for the Eurasian kind. You can't fool nature.

Similar in intent to the Pit and Mounds Project is another ecological restoration being undertaken on the eastern side of Rice Lake, Ontario, where members of Alderville First Nation, under the coordination of Rick Beaver, have parcelled off and protected a forty-four-hectare chunk of tallgrass prairie – the largest remnant of it in Central Ontario – to create a natural preserve called the Black Oak Savannah/ Tallgrass Prairie. Black oaks (*Quercus velutina*) are notoriously difficult to grow. A friend gave me three carefully nurtured seedlings last summer, and I planted them with great ceremony beside the vegetable garden in the fall; none of them made it through the winter, and for once it wasn't because of the mice. They just seemed to give up. In Toronto's High Park, a number of huge, hundred-year-old black oaks sit despondently on their hilltop above Grenadier Pond like gigantic bull elephants that have failed to produce offspring. Donald Culross Peattie, in his *Natural History of Trees*, remarks that black oak is not a favourite for planting, being "too heavy in form, too narrow in the crown, too unkempt in its winter outline," to which I would add, too likely to die in captivity. This may have saved it in the wild. Those in the Black Oak Savannah / Tallgrass Prairie seem to be doing just fine. Rick Beaver, who is an accomplished artist when he is not coordinating restoration projects, points out that the Ojibway used the inner orange bark to make yellow dye, and boiled out the tannin for use as an antiseptic, a tonic, and an emetic. Infusions were also given to treat asthma, and the acorns were boiled and eaten.

But the Alderville First Nation project is more than a simple heritage site, more than a wall of legislation around a habitat that is then left untouched: it is an active garden. Native species are planted – Prairie lily, wild bergamot, New Jersey tea, chokecherry, and buffalo berry, as well as white birch, red pine, and black cherry – and alien invaders are eradicated as much as possible. These include black medic, spotted knapweed, Canada thistle, mullein, king devil, Queen Anne's lace, and red clover, all of which came to the New

World mixed with grain seeds (which were also exotic) or, in many cases, lodged in the boot soles or pant cuffs of settlers, which gives new meaning to settlers' other name, "planters." In most places, introduced species outnumber and outcompete native plants. A survey of Toronto's Leslie Street Spit – a landfill site stretching five kilometres into Lake Ontario and left to run wild – turned up 278 species of plants, only 122 of which were native to Southern Ontario.

The Alderville project is also trying to restore farmland to tallgrass prairie. In 2002, volunteers reclaimed an oat field by planting six species of grasses and twenty-three species of wildflowers, using seed collected from the tallgrass-prairie section of the preserve. Also that year, says Beaver, "an army of volunteers planted 2,348 Blue lupine seed plugs germinated from seed collected in the vicinity." Blue lupine (*Lupinus perennis*) is the principle food- and egg-site of the Karner blue butterfly, a species that has been extirpated from Canada since the late 1980s. Once the lupine habitat has stabilized, the project plans to reintroduce captive-bred Karner blues, "and," says Beaver, "what a colourful day that will be!"

Planting a garden to make a wilderness sounds like a contradiction in terms. The idea that unifies wilderness and garden is that in either one we can climb out of our mundanity and renew our connection with the sublime. In *The Conquest of Paradise*, Sale writes that "in its attitude to the land, and the creatures thereof, a culture reveals the truest part of its soul." He notes that when the first Europeans came to North America they had a chance to leave behind the attitudes that had turned Europe into a largely human-made artifact; we could, he laments, have started afresh, learned from the people who were already here and entered into a new relationship with the natural world. We didn't do it. We chopped and shot and ploughed what we found, and replaced it with our old familiar European stuff. Turning it back into wilderness, or at least as close to wilderness as we can manage, does not, in comparison, seem such a huge step.

Ecological restoration – wilderness gardening, using our knowledge of nature to heal and restore rather than to raze and flatten – would go some way toward healing the rift between ourselves and nature. Galileo was a devoted gardener; in his walled plot behind his various households he always maintained grapevines and citrus trees. He grew chartreuse citrons in terra cotta pots and sent them to his daughter, Suor Marie Celestre, in her convent in San Matteo, where she candied them and sent them back to her father. Perhaps it was his propagation experiments with fruits and vegetables that convinced him of the power of change. Like Karen Landman and Rick Beaver, he believed that Nature could be renewed.

"For my part I consider the Earth very noble and admirable precisely because of the diverse alterations, changes, generations, etc. that occur in it incessantly," he wrote in his Dialogue between the proponents of Aristotelian immutability and Copernican evolution. The speaker is Sagredo, a Copernican. "If, not being subject to any changes," Sagredo continues, "it were a vast desert of sand or a mountain of jasper, or if at the time of the flood the waters which covered it had frozen, and it had remained an enormous globe of ice where nothing was ever born or ever altered or changed, I should deem it a useless lump in the universe, devoid of activity and, in a word, superfluous and essentially nonexistent."

As it is, the Earth is capable of infinite change, if change is allowed to take place slowly enough; in time, it can even change back into what it was. Despite our efforts to render it into a vast desert of sand (global warming) or an enormous globe of ice (global cooling), it remains noble and admirable. It isn't Paradise, perhaps, but neither is it a useless lump.

Bringing Back the Dodo

The call of the dufuflu bird
For which I have an ear
Falls like the uncreating word,
But only some can hear.

— GEORGE JOHNSTON, "The Dufuflu Bird"

As a science-fiction writer, Isaac Asimov looked into the future, but as a scientist, he also had a healthy respect for the past. In 1962, in a book called *The Genetic Code* – a history of the discovery of DNA and its role in our understanding of how life works – his Janus-like gaze combined the two perspectives. He considered a number of past scientific discoveries, traced their expansion into everyday life, and concluded that science works its way into society very quickly indeed. "Sixty years," he wrote, "seems to be the typical time interval from scientific breakthrough to full flower." For example, the Danish physicist Hans Christian Oersted discovered electromagnetism in 1820, and by 1880 we had the incandescent light bulb; Thomas Edison noted the transference of electrical energy from a filament to a plate in 1883, and televisions and computers were on the scene in the early 1940s.

In 1944, he went on, the Canadian-born biologist Oswald T. Avery, working in New York's Rockefeller Institute Hospital, isolated a substance that appeared to be capable of changing one strain of bacteria into another. The substance was deoxyribonucleic acid, better known as DNA. Avery didn't exactly discover DNA — he was following up on a series of experiments conducted in the 1920s by British pathologist Frederick Griffith — but he was the first to suggest that DNA (not protein, as had previously been believed) was the active factor in inherited variation between organisms, and was therefore the foundation of all life. That discovery changed our perception of life from a force that seemed bent on controlling us, into a combination of molecules that could be manipulated to suit our needs. However we chose to exercise that control, Asimov predicted a major genetic breakthrough sixty years after Avery's initial discovery. "I feel confident," Asimov wrote, "that, if we survive, the year 2004 will see molecular biology introducing triumphs that can now barely be imagined."

Let's assume that Asimov was approximately right — scientists like to work with a certain amount of wiggle room, which they call "correction factors" — and that a big breakthrough in genetic science is about to happen within the next few years (James Watson and Francis Crick twigged to the double-helix structure of DNA in 1953, so we can give Asimov another decade). What form would these triumphs take? The final mapping of the human genome comes to mind, but that seems more a stage in a developing triumph than a paradigm shift on its own. It has already led to attempts at such a shift: geneticists now talk about shaping the genetic makeup of our offspring by first identifying and then manipulating the specific genes responsible for such "desirable" but undefinable traits as beauty, intelligence, or athleticism. Called "germline therapy," this is merely the reappearance in a modern lab coat of the old, discredited notion of eugenics, first proposed in the late nineteenth century by Francis Galton and carried out with horrific assiduousness in Nazi Germany. Though germline therapy would count as a genetic breakthrough, so many people are against it – including the Pope and the U.S. president – that it is unlikely to happen within Asimov's time frame. As Everett Mendelsohn, Harvard's professor of science history, has recently noted, "The genetic revolution begun in 1953 is unlikely to have run its course, either scientifically or culturally, by the time our grandchildren celebrate its one-hundredth birthday, in 2053."

Cloning, then? Surely cloning is very old hat, at least the cloning of animals. The first clone – a frog – was produced as early as 1951, seven years, not sixty, after Avery's groundbreaking work. The frog was produced when the nucleus from one frog cell was transferred into an egg taken from another frog from which the nucleus had been removed. Reinserted into the female, this renucleated egg then divided in the normal way until a tadpole was produced. The scientists who conducted this experiment attempted 197 nucleus transplants, from which they created twenty-seven embryos that went on to survive the tadpole stage. This was an incredibly high success rate; forty years later, the scientists who produced Dolly the Sheep made thousands of attempts before coming up with a single successful clone. Since Dolly, however, genotechnicians have cloned a number of animals, including pigs, goats, cows, mice, and white-tailed deer. And cloning endangered species has also been done – a gaur, a rare species of Asian ox, was born to a foster cow named Bessie in Ohio, in January 2001; and an African cat, now in the Audubon Zoo in New Orleans.

What, then, is left for a breakthrough? One possible candidate on the horizon is the cloning back into existence of an extinct species.

Jurassic Park prepared us for it. The book on which the movie was based was fiction, but like Asimov, Michael Crichton grounds his fiction on fairly sound scientific principles; he takes plausible if controversial hypotheses and presents them as though they were fact. At the time *Jurassic Park* was written, it was not generally accepted that dinosaurs travelled in vast herds, laid eggs in huge colonies, and cared for their young like birds, as Crichton had them doing. Those theories are considered state-of-the-science today. It was also hypothetically possible to salvage dinosaur DNA from some ancient saurian gene pool, but no such pool had been found. Crichton imagined dinosaur blood in the bodies of mosquitoes trapped in amber, which is 80-million-year-old fossilized tree resin. Scientists have rescued fragmentary DNA from such sources – UCLA researcher George Poinar has rescued fragments of

125-million-year-old beetle DNA from a chunk of Lebanese amber – but so far no dinosaur blood. Dinosaur material might one day be pulled up from the Alberta tar sands, preserved in oil rather than fossilized. Burnable logs from Cretaceous trees have been retrieved in that fashion, so why not a hadrosaur haunch? But the vast interval between the age of the dinosaurs and today makes the likelihood of obtaining living DNA even from well-preserved dinosaur flesh extremely remote.

Something more recently extinct, perhaps? In 1997, a family in northern Siberia discovered a woolly mammoth that had cloners rubbing their hands in anticipation. The mammoth was encased in glacial ice off the coast of the Kara Sea. The top of its cranium was exposed and most of its brain was missing, but other than that it was as intact as on the day, 20,380 years ago, when it was flash frozen. Woolly mammoths (*Mammuthus primigenus*), direct ancestors of today's Asian and African elephants, roamed the earth from 4 million years ago until the Wisconsin Ice Age, which ended eleven thousand years ago, and bits and pieces of them have been turning up across the Arctic for years. There are an estimated 10 million specimens suspected to be buried in the Siberian permafrost, and more in North America. I know a retired gold miner in Dawson City, Yukon, who pulled dozens of mammoth tusks, teeth, and bones from his placer mine in the permafrost. The better specimens he donated to the Museum of Natural History in Ottawa, the rest he laid out neatly for me to see on his kitchen table, along with bones from extinct reindeer and bison species. Fascinating though his collection was, the stuff looked pretty dead to me, and I doubt any useful DNA could be extracted from it.

But Zharkov, as the Siberian woolly mammoth was named, was practically fresh meat, although admittedly past its best-before date. Bernard Buigues, the Dutch scientist who rushed to the discovery site, carved the creature out of the glacier, leaving it in a twenty-three-ton block of ice, transported it 250 kilometres to the city of Khatanga, and began thawing parts of it with a blow-dryer. The thick tendrils of metre-long hair felt vibrant and alive, he reported, and the chamber he was working in filled with "the pervasive, unforgettable smell of elephant." Dick Moll, one of Buigues's team members, called the discovery "a dream come true."

The big question, he said, was whether the carcass would yield living DNA that would allow the beast to be cloned. He was hopeful that it would. "What we need to find," he said, "is DNA in parts of the marrow, in big bones like the femur, or in the soft organs."

After a year of blow-drying, however, the possibility of cloning a woolly mammoth from Zharkov was declared "unlikely" by biologist Alexei Tikhonov at the Zoological Institute of St. Petersburg. Water damage, freeze-thaw cycles, ultraviolet radiation, and chemical decay had all taken their toll on the DNA's fragile cell walls. Although a partial genetic code was extracted, enabling scientists to establish more detailed taxonomical links between woolly mammoths and modern elephants and telling them much about why the species disappeared so suddenly – their population nose-dived in the last thousand years of their span, probably because of a herpes-like disease still carried by Indian elephants and deadly to the African variety – there is not likely to be a herd of woolly mammoths meandering through Tombstone National Park next year. (A team of Japanese scientists, undeterred by the Zharkov's failure, is still trying to extract DNA from a woolly mammoth found in 1994, with the idea of impregnating an Indian elephant with enhanced mammoth sperm. If successful, they plan to exhibit the result – which would be 88-per-cent mammoth – at an Ice Age theme park being built near the River Kolyma in Siberia.)

Since DNA ceases to repair itself after death and deteriorates over time, scientists are looking for a species that vanished more recently than woolly mammoths. The list of candidates is long and is growing rapidly, thanks to us. Extinctions have been escalating arithmetically since the Industrial Revolution. Between 1800 and 1850, only two species of mammals are known to have disappeared from human causes – the eastern bison in North America and the Hispaniola hutia, a West Indian rodent. From 1851 to 1900, thirty-one mammals went extinct, including the intriguing South African zebra known as the quagga. From 1901 to 1944, forty more vanished, among them the Barbary lion, the Japanese wolf, and the Texas grizzly. At present there are more than six hundred endangered mammal species worldwide, and innumerable bird, fish, insect, and plant species. In a few years, cloners will be able to take their pick.

In fact, at least two projects involving recently extinct species are currently underway. Scientists are busily trying to resurrect the burcado, an extinct Spanish mountain goat, and the thylacine, or Tasmanian tiger. Living tissue was taken from the last remaining burcado a year before it died (she was called Celia; a tree fell on her in 1998), but it is not yet known whether enough DNA can be extracted from it to attempt cloning. Ditto the Tasmanian tiger. Although rumours of its continued existence in remote parts of Tasmania persist (like those of the eastern cougar), Australian geneticists are pinning their hopes on a fetus that was preserved in alcohol in 1866. They expect it will take at least ten years to find out if the fetus has usable DNA, but "if they succeed," *Globe and Mail* science reporter Anne McIlroy recently observed, "the term 'going the way of the dodo' will have to be redefined."

"Dead as a dodo" means of course a thing so dead it doesn't bear thinking about. But surely nothing is ever that dead. Not even the dodo. In fact, from a purely symbolic standpoint, the dodo (*Raphus cucullatus*) has a claim to being the species most worthy of resurrection, and on the face of it is at least as promising a candidate as the burcado or the Tasmanian tiger.

Related to the dove and native to the three Mascerene Islands in the Indian Ocean off Madagascar, the large, flightless, dumb-as-a-post bird (the name "dodo" probably comes from the Portuguese word *doudo*, meaning "simpleton") was discovered by Europeans in 1598, when a ship under the command of Dutch admiral Jacob Cornelius van Neck landed on Mauritius, the largest of the Mascerenes, to take on water. Dodos covered the island. Van Neck called them *walghvogels*, "disgusting birds," because, he wrote, "the longer they were cooked, the less soft and more inedible their flesh became."

Inedible or not, dodo meat was a staple of East India Company crews for the next four decades. The Dutch planted a colony on Mauritius, and colonists not only killed dodos with abandon and sold their salted flesh to transient merchantmen, but also introduced pigs and monkeys, which feasted on dodos and their eggs. The carnage took its inevitable toll. In 1638, an East India Company captain named Peter Mundy reported that, whereas on the previous year's outing he had seen a few dodos on Mauritius, on this return voyage he "now mett with None." A few were sighted (and killed) by a shipwrecked Dutch sailor in 1662, and that was that.

As naturalist David Quammen has pointed out in *The Song of the Dodo*, *Raphus cucullatus* enjoys the dubious distinction of being the first creature to become extinct as a direct result of human intervention: it is, he writes, "the sort of legendary bird of extinction," and our realization that we played a part in its demise represents "a very, very important time in the dawning of human consciousness." We would be undoing a great wrong if we could bring it back to life; we would also be giving hope that other great wrongs we have committed in our short span on this planet may be reversed. Isaac Asimov would approve. All we need is a little DNA.

But there's the rub. At the time of its extinction, there were only two stuffed dodos in all of Europe. Both belonged to John Tradescant, an avid collector of ornithological curiosities. Upon his death in 1683, Tradescant's dodos were bequeathed to Oxford University's Ashmolean Museum, where they gathered dust until, in 1755, the museum's directors grew tired of them and decided to do some housecleaning.

Before the carcasses were tossed into the incinerator, however, some prescient staffer removed the head and one foot from one of them, and that scanty material represents the best possible source of dodo DNA for today's geneticists. Dodo bones, dug up from a riverbed in Mauritius in 1863 and now in the American Museum of Natural History and the Smithsonian Institution, are other possible sources. There is also the head of a dodo in the Museum of Natural History at the University of Florence. I've seen it. It looks like it was carved out of grey chalk. It looks deader than the mammoth bones I saw in Dawson City. In 1961, the eminent Yale biologist G.E. Hutchinson speculated that "it is just possible that some of the fossil bones will be found to contain enough protein for an immunological test of the possible affinities [of the dodo] in the future, though this obviously should not be done on such relatively rare material until the technical details are rather better developed than they appear to be at this moment." Technical details are now better developed – a Netherlands research team recently discovered a cache of some seven hundred dodo bones in a swamp on Mauritius, and although they will help establish the dodo's affinity with the pigeon, they are not expected to yield any usable DNA.

There is, however, another candidate for resurrection, one whose bones are a tad fresher. It is, moreover, closer to home: the extinct eastern North American shorebird, the great auk. A relative of puffins and guillemots but much, much bigger, the great auk (*Alca impennis*) was curiously like the dodo in many respects. It was flightless, clumsy on land, notoriously easy to kill, and it inhabited islands. Normally solitary, during breeding season it colonized rocky islands off the coasts of Labrador, Newfoundland, Greenland, and Iceland in the millions. Sailors armed with nothing more than clubs and gunny sacks could kill hundreds in an hour. It was a black bird with a white front and dark brown head, with white cheek patches between the eye and the beak – the Gaelic name for it was *gaerrabhul*, meaning "strong stout bird with spot," which is possibly why the Vikings called them *geirfugl* and the later English fishermen knew them as garefowl. The Celts called them Pen-gwyns, or "white heads," which was the name Sir Francis Drake later transferred to the southern seabird now known as the penguin.

Numerous early reports tell of killing sprees on great-auk breeding islands; ships' holds would be filled with barrels of salted auk flesh and eggs cushioned in sea grass. Apparently, auk tasted better than dodo. By the year 1000 they were no longer common on the European side of the Atlantic, but in 1534, John Cabot described vast colonies of great auks on Newfoundland's Funk Island (whence, probably, its name: "The smell of Funk Island is the smell of death," writes Franklin Russell), which now supports a million murres. The auks are long gone. They continued to decline throughout the seventeenth century, when the feather trade demanded their down for beds and their oil for lamps and stoves. By 1821, they were gone from Newfoundland; in 1830, a volcano off the coast of Iceland sank their island refuge there, and a mere fifty of them made it to the nearby island of Eldey. When museums around the world realized there were only a handful of great auks left, they offered money for specimens, and the fifty remaining great auks were killed one by one for their skins. The famous Audubon painting of a great auk was done from a stuffed skin; he never saw a live bird. In 1844, there were only two left. On June 4 of that year, three Icelandic fishermen rowed out to Eldey Island and killed them.

In 1863, however, a guano miner on Newfoundland's Penguin Island uncovered a cache of frozen great auk carcasses buried in birdlime, and these were promptly distributed to museums and scientific institutions around the world. There are now more than eighty stuffed great auks, as well as numerous skeletons and eggs, from which DNA might profitably be extracted. The great auk, already the Canadian dodo, has much potential for becoming the Canadian Lazarus.

The dodo is an unlikely candidate for resurrection for another reason: there is the problem of the host female. What species would bear the embryo that would grow into the cloned resurrectant? Cloning works because the renucleated egg is placed in the womb of a receptive host, and that host should ideally be a member of the same family, if possible the same genus, as the donor. When recent attempts were made to clone the endangered African wildcat (*Felis sylvestris lybica*), the renucleated egg was borne by a domestic cat (*Felis sylvestris catus*). If Tasmanian tigers (*Thylacinus cynocephalus*), which are carnivorous marsupials, were to be brought back, it would be through the birth canals of its nearest relatives, Tasmanian devils (*Sarcophilus harrisii*). Woolly mammoths, should they surprise us, could be borne by living African or Indian elephants. But what would serve as a surrogate mother for the poor dodo? The only other living members of the Raphidae (a subfamily of the Columbidae, which includes pigeons) are the solitaires, and they are becoming extremely rare. Quammen notes that Réunion and Rodrigues islands each "harboured a large, flightless species of solitaire (*Ornithaptera solitaria* on Réunion, *Pezophaps solitaria* on Rodrigues) that, like the dodo, had pigeon affinities," but both are extinct. The Townsend solitaire (*Myadestes townsendi*) is a North American thrush, and definitely not a candidate. The next best bet is probably the pink pigeon (*Nesoeanas mayeri*), which is still found on Mauritius but in very small numbers; it is itself on the verge of extinction. Resurrectionists might be faced with having to clone a number of pink pigeons in order to secure enough surrogate mothers to then try to clone a dodo. (Some ornithologists hold that the dodo was more closely related to rails than to pigeons, in which case the lineup of possible surrogate mothers lengthens considerably. New Zealand's *Notornis* is the world's largest living flightless rail and, like the dodo, lays a single big, white egg – unlike pigeons, which lay several small, mottled eggs.)

The great auk also has living relatives, among them the razorbill, or razor-billed auk (*Alca torda*). Although smaller in size, it is at least of the same family, the Alcidae, as are murres and puffins. In a pinch, one of the Antarctic penguins might also serve, though they only stand and wait.

A question more urgent than whether we can clone an extinct species is being asked by many environmentalists. Should we try at all to bring back species that have long vanished? Or even species on the point of vanishing? Nice as it would be to see what a dodo or a great auk really looked like, and to stop having to worry about burrowing owls or red-sided daice, the issue is not without controversy. Saving the endangered California condor by capturing the remaining individuals and breeding them in the San Diego Zoo was opposed by environmental guru David Brower, who thought a dwindling species should be allowed to "die with dignity," and did not want to see the last individuals of a once noble species hunched over in a zoo with radio transmitters implanted in their anuses being taught how to eat by a hand puppet. In the end, the condor was captive-bred, zoo-raised individuals were released into the wild, the species was saved − and its original ranges are still being swallowed up by development. The California condor flies again, but hikers have a better chance of seeing one in Arizona's Paria Canyon–Vermilion Cliffs Wilderness than they do in its native California.

There is also an ecological objection to resurrecting vanished species. We all know nature's attitude to a vacuum. When a species goes extinct, its niche is promptly filled by other species. Where the great auk once fished and bred, murres, razorbills, puffins, and guillemots now thrive; bringing back the great auk would be akin to introducing a new, competing species. The great auk could become the European starling of the ocean. Once again, we would be tampering with an ecosystem about which we know too little.

Clones represent the ultimate loss of biodiversity. Having breakfast recently in the dining room of a rural hotel, I found myself being impassively contemplated by the decor: the mounted heads of half a dozen white-tailed deer, staring down at me in glass-eyed uniformity. What struck me as I munched my GMO cornflakes was that all six heads were identical. All had no doubt been shot locally, and so all had belonged to a fairly restricted gene pool; each head on the wall was probably closely related to the others. In a similar population of cloned animals, each would be genetically identical to the others: if one came down with some inheritable disease, they all would. Every clone has the same genetic makeup as its nucleus-donor. One reason humans have survived this long is that we keep increasing our biodiversity through interbreeding among distant populations; when something like bubonic plague or Asian flu sweeps through, enough of us are resistant to it to perpetuate the species. That wouldn't happen in a cloned community. If the cell donor was susceptible to heart disease, or happened to be ill-adapted to its environment, the entire population would be in trouble.

Perhaps we'd better hope that Isaac Asimov was wrong, after all. Perhaps we don't need a breakthrough in genetic research. It's hard to see the cloning of animal species as anything but practice for more controversial needlework – the cloning of humans, for example – and therefore may be a road we should not travel. Asimov warned that "this new understanding might be abused, might serve as the source of a new horror: the scientific control of life." Dolly died on February 14, 2003 – Valentine's Day – and since she was created by the Roslin Institute of Edinburgh, her body has been donated to the

National Museum of Scotland, where it will be stuffed and put on display. Maybe it should be placed beside the skeleton of a dodo, along with the bodies of great auks, passenger pigeons, South African quaggas, and other extinct creatures, all housed in their own gigantic edifice: the Museum of Lessons Unlearned.

ATWOOD AND MCKIBBEN

Т he launch party for Margaret Atwood's novel *Oryx and Crake* was held in a half-demolished hotel in Toronto's west end – the exposed lathwork and crumbling bricks seemed a fitting backdrop for Atwood's apocalyptic vision of a crumbled world given over to genetic engineering gone terribly wrong. During her remarks to the assembled literati, Atwood mentioned that she was reading Bill McKibben's book *Enough* and found it, she said, "a sort of non-fiction companion piece to my novel." Indeed, the American nature writer's depiction of a world obsessed with genetic engineering and nanotechnology is as brilliant as Atwood's, and no less hair-raising for being a work of non-fiction. It is therefore doubly disturbing to consider the two

174

works as two sides of one coin – to read them simultaneously, as it were, with *Enough* in one hand and *Oryx and Crake* in the other.

McKibben has been a prophet in his own country since he published *The End of Nature* in 1989. In that book he suggested we can no longer support the environmental and psychological costs of the fossil-fuel industry, which entailed the degradation of nature as a corollary of seeing the world as an endlessly violatable source of raw materials. In *Enough*, he focuses on another clear and present danger: that of living in a world permanently altered by genetic manipulation, a world in which not only our food and pets are altered at the cellular level, but also ourselves. "The genetic modification of humans is not only possible," he writes, "it's coming fast; a mix of technical progress and shifting mood means it could easily happen within the next few years." This sounds a lot like the "triumph" in genetic science that Isaac Asimov warned us about: the complete control of the human environment achieved by the complete control of human beings.

We know that Monsanto's bioengineers can already insert DNA from soil bacteria into a canola plant to "improve" the species by making it resistant to certain broad-leaf herbicides. They can genetically modify potatoes to make them poisonous to spiders. Scientists have cloned sheep, rabbits, mice, and pigs, and implanted rabbits with jellyfish genes to make them glow in the dark (yes, really). They have made worms live seven times longer than their natural lifespan. Do we imagine, McKibben asks, that they will stop at playing around with rabbits and worms? Do we think that experiments in "improving" human beings, by making us live longer, healthier, and smarter, are not being carried out in labs around the world? As geneticists are finding out, our genes control almost everything about us, and geneticists can control our genes. If Atwood alarms us with her depiction of the nightmare that biotechnology can visit upon the world, McKibben horrifies us with the warning that, if we close our eyes to it, that nightmare will unfold – in fact, is unfolding as we read.

In *The Future of Technology*, a book similar to McKibben's written by German scientist Friedrich Georg Juenger in 1936 (but not published even in Germany until 1946 because of its anti-progressive views), Juenger maintains that Utopian writers such as Jules Verne or the American novelist Edward Bellamy are not prophets or visionaries; they do not look into the future and see things that cannot be posited by looking at the present. "What they project into the future," Juenger writes, "is merely a possibility emerging in the present, expanded by them in a logical and rational manner." That is what makes Utopian literature so charming, and Dystopias so chilling. *Oryx and Crake* is speculative fiction, that is, futuristic, in the same sense that George Orwell's *1984*, or Aldous Huxley's *Brave New World*, were. All three novels are examples of *vaticinia ex eventu*, history disguised as prophecy. Orwell chose the date 1984 by simply reversing the last two digits of the year in which he wrote the book, the implication being that the horrors he described taking place in the future were at the very least logical extensions of patterns he discerned in contemporary life. The Soviet novelist Mikhail Bulgakov performed similar sleight-of-hand in *The Master and Margarita*, a work of fantasy in which the Master, Beelzebub, was a thinly disguised Josef Stalin. *Oryx and Crake*, while it pretends to take place in some distant future, works so well upon our imagination because it contains elements we know are present among us now. "As with *The Handmaid's Tale*," Atwood explains, *Oryx and Crake* "invents nothing we haven't already invented or started out to invent. Every novel begins with a *what if*, and then sets forth its axioms. The *what if* of *Oryx and Crake* is simply, *What if we continue down the road we're already on?*"

In the novel, the woods are populated by escaped trans-genic animals (having the spliced genes of two species, known in the trade as "chimeras") that have gone feral — rakunks (raccoons and skunks), snats (snakes and rats), wolvogs (wolves and dogs), and pigoons (pigs with human tissue, created to supply hospitals with transplant organs — five kidneys in a single pig). Of course our experiments are going to get loose — look what happens in *Frankenstein*, or *Dr. Jekyll and Mr. Hyde*, or *Jurassic Park*. Also released into the wild is a new race of human beings, created by splicing desired traits from human and nonhuman donors into human embryos. This was Crake's project, and the people he created are called Crakers. Crakers are top-of-the-line chimeras. They have scent glands from various animals, bobkittens or rakunks, so that their urine contains pheremones that warn other animals away; they are self-healing, purring like cats at the same frequency as bone-mending ultrasound; their children mature in five years; they are vegetarian, living like sheep or deer on leaves and grass; their sweat glands secrete a citrus-like compound that wards off mosquitoes. We may think such notions are farfetched, but it was recently announced in a scientific journal that Chinese scientists have succeeded in creating a human-rabbit embryo — a rabbit with human tissue, ideal for testing certain cancer drugs, no doubt, but what might Atwood call such a creature? A hubbit?

In McKibben's view, we are already proceeding down Atwood's road. Science, he writes, can now "do to humans what we have already done to salmon and wheat, pine trees and tomatoes. That is, to make them better in some way." By deleting, modifying, or adding genes in developing human embryos, scientists can make us "taller and more muscular, or smarter and less aggressive, maybe handsome and possibly straight, perhaps sweet. Even happy." It is, McKibben admits, "in certain ways, a deeply attractive picture." No need for Medicare, no relying on Prozac or Paxil or Botox. Everyone is born perfect and stays that way. Forever. As soon as we get past certain ethical stumbling blocks to stem-cell research (the proper cells for cloning new human organs have to be taken from aborted or miscarried fetuses), which so far has been focussed on treating diseases, scientists may begin to replace worn-out or defective body parts — hearts, kidneys, hips — with brand-new ones grown *in vitro*, eventually creating an entirely new person as each part of the old one disintegrates and is replaced. O, brave new world, that hath such people in it.

Perhaps we look upon such innovations with equanimity, at least at first thought, because we know that throughout history someone has always been trying to make human beings "better" – through prayer or confession or analysis – without having had much effect on our day-to-day lives. We have become wildly enthusiastic, or mildly tolerant, or even faintly amused, by such attempts, and have then gone bungling along in the usual fashion. But when Religion or Psychology fail, as they mostly have, to improve us by altering our collective behaviour, the results do not entail the end of civilization as we know it. Their failure does not threaten our very survival, at least not directly.

But when Science fails, it fails on a massive scale. The atomic bomb. Bhopal. Chernobyl. Atwood shows that the attempt by scientists to make us a better species is a deeply flawed and disturbing one. McKibben agrees. In a world inhabited by "posthumans" or "transhumans," he writes, in which everyone lives exactly according to their design specs, "there won't be moral decisions, only strategic ones." We won't ask, Is it right to ensure that my baby has an IQ of 180? We will ask, Is 180 enough? What if every other parent in the subdivision soups up their kids to 220? Deciding not to soup up our own kids' IQs, or not to remove genes that can cause cystic fibrosis or breast cancer, or not to make them blond, blue-eyed, and svelde, "could come to seem like child abuse."

"The tide of human desire," writes Atwood, "the desire for more and better, would overwhelm them. It would take control and drive events, as it had in every large change throughout history." In Atwood's world, Crakers are perfectly formed, the women are universally beautiful, and the men invariably athletic – great gene-pool specimens, if you don't care too much for independent thinkers. They are disease-free. There is no violence among them. The females come into heat for a day or two once a year, "as did most mammals other than man," and mate with whichever male or males are near to hand. There is no sexual jealousy, as there is always a female in heat somewhere and a male to service her; the Crakers feel no desire that cannot be immediately gratified. Most of the time they feel no desire at all.

The problem, however, doesn't come from the docile Crakers. They seem content enough. When trouble comes, it comes from those who have not been selected for improvement, the old-style humans who have been abandoned to live among the rakunks, snats, and wolvogs. They live in Pleebland, where their chief function is to provide DNA traits that can be harvested and used by the transgeneticists to create evermore perfect Crakers. "People come here from all over the world," Crake tells the narrator on a foray into Pleebland, "they shop around. Gender, sexual orientation, height, colour of skin and eyes – it's all on order, it can be done or redone. You have no idea how much money changes hands on this one street alone."

"If the technology is going to be stopped," McKibben writes, "it will have to happen now, before it's quite begun." And he devotes two chapters to what we must do to avert the nightmare envisioned by Atwood. Science, he says, has one vision for the future – a future of enhanced humans who no longer suffer from hereditary conditions such as cancer or alcoholism, who are designed to be smarter, to run faster, to think and behave according to certain pre-set patterns. Politicians and corporations approve of that vision because it means a society of uncritical thinkers and complacent consumers. We need, he says, to counter that future with an equally compelling vision of our own, "with some other account of who we are, and what we might be."

For a start, we "need to survey the world we now inhabit and proclaim it good. Good enough." No more griping. In *Hope, Human and Wild*, McKibben countered some of the pessimism of his first book by looking around the globe and finding places where the environment has not yet been destroyed or degraded, places where citizens have proclaimed a halt to unsustainable development and begun healing the Earth. We need to do more of that, McKibben says, and we need to do it fast. If we can see the present as sufficient for our needs, if we can agree that we value the struggle we now engage in to make ourselves healthier or smarter by natural means, then "perhaps we can figure out how to avoid these new technologies and the risks – physical and existential – that they pose."

Liking ourselves as we are doesn't seem to be much of an arsenal against "the technological momentum" McKibben and Atwood describe in such chilling detail. McKibben admits that it hasn't worked all that well in the past. "We may keep our TVs in the closet," he says, "but we still all live in a TV society," and we already depend to an enormous degree on the very technologies McKibben says we have to scrap if we are going to maintain any kind of human dignity.

Atwood agrees. Her narrator, a non-Craker named Snowman, likes himself pretty much as he is. The novel comes close to suggesting that the world can be saved through some kind of religious reawakening. In the old battle between Science and Religion, Science has dominated for at least the past four centuries, but these things are cyclical; Religion held sway during what we, living in the scientific Age of Enlightenment, refer to as the Dark Ages, and perhaps now is a good time for Religion to make a comeback. In *Oryx and Crake*, the transgeneticists are unable to sever a few crucial human characteristics in the Crakers. Crake "hadn't been able to eliminate dreams," one of the Crakers tells Snowman. "We're hard-wired for dreams, he'd said. He couldn't get rid of the singing, either. We're hard-wired for singing. Singing and dreams were entwined." One of the Crakers' recurring dreams is to discover who made them. They compose songs to that unknown creator.

It may be worth recalling here that the first *vaticinium ex eventu*, the earliest-known work of history disguised as futuristic prophecy, was the Book of Revelations. Composed between 95 and 97 AD, this ultimate chapter of the Bible has been read as a vision of the second coming of Christ in the new millennium, when the trumpets will sound on the Day of Judgment, and all true believers will rise from their graves and follow Him into His promised kingdom. But two millennia and no second coming later, interpretations of the Book have swung from the ecstatic to the mundane: the work is now thought to be a kind of *roman à clef* of the days immediately preceding St. John's vision of the Apocalypse, a kind of clandestine history, full of hidden contemporary references — Nero as the Beast, the fall of the Roman Empire, and so on – the way Orwell wrote *1984*, or Mikhail Bulgakov, with Soviet censors peering over his shoulder, wrote *The Master and Margarita*.

Oryx and Crake doesn't end on a religious note, but rather with the possibility that a pocket of non-Crakers, flawed though they be, has survived the cataclysm. *Enough* also goes out opting for the here and now. Indeed, the whole point of "enough" is rejecting the notion of "more" or "better," which is what both Religion and Science invariably offer, in favour of accepting what and who we are, warts and all. McKibben revels in the mundane, but his is an elevated mundanity. Call it humanity. "Environmentalists, and I am one," he writes toward the end of the book, "have always been concerned with keeping the wonder of the present moment alive. With valuing what *is*."

Ironically, keeping the wonder alive is a religious impulse, while valuing what *is* is a scientific one. But they are both pure impulses – purely religious and purely scientific – not poorly conceived Utopian schemes to make the world better for a select few. Not applied religion, which is dogma, or applied science, which is technology. Both Atwood and McKibben warn against the passing of the natural world, one that contains pain and suffering and cruelty as well as joy and kindness and rapture, the whole Pandora's box full of tricks. Without them, there is no need for hope.

SEND IN THE CLONES

My neighbour, who lives down the road from where I live in the country, owns a dog, which he keeps tied up in a fenced-in compound behind his house. The dog is fed and watered regularly, but apart from that receives very little attention from its owner. As a result, it barks constantly, it is barking now, hopeless howls for affection that will go on all day and most of the night. For an hour in the evening, however, when the coyotes in the woodlot across the road begin their wild ululations, the dog will suddenly shut up and hide in its kennel.

As far as our relationship with nature is concerned, we are that dog.

The first scientist to propose that human beings are, like dogs, a domesticated species was Johann Friedrich Blumenbach, a German anatomist who in 1776 published *On the Natural Variety of Mankind*, the results of his lifelong study of human skulls. It was Blumenbach who divided the species *Homo sapiens* into five distinct "races" – Caucasian, Amerindian, Malaysian, Mongoloid, and Ethiopian – based on careful measurements of the many mandibles, orbits, and braincases he had collected from around the world. These are still more or less the five races recognized by most workers in that field, and Blumenbach is credited with being the father of anthropology: he was the first scientist to look upon human beings as a species like any other. If he had left it at that, his reputation might have remained obscure but untarnished, but having made his list of races he then felt compelled to put it into some kind of order. Which of the five groupings of humankind was the type specimen? he asked. Which was ancestral to the others? Since he had to base his ranking on something, he had the unhappy notion that Caucasians (those fair-skinned, pleasant-featured Georgians who live on the eastern slopes of Mount Caucasus) were the "most beautiful" of all humans, and he situated each race in his list according to the degree to which it seemed to have "degenerated" from this type. For this he is also known as the father of modern racism.

Since he was writing eighty-four years before the publication of Darwin's *Origin of Species*, Blumenbach knew little about the forces that drove evolution, but he did know a great deal about the breeding of domestic animals – a continuing process among "civilized" races for about ten thousand years.

He knew that all domestic dogs, for example, no matter what colour, temperament, or skull size, were members of the same species, *Canis familiaris*; ditto all domestic cats, horses, cattle, pigs, goats, sheep, and pigeons. He also knew that one definition of a species was that any fertile male and female within it, regardless of "race," could mate and produce viable offspring. A Highland bull could impregnate a Holstein cow and produce a calf, and a Mexican chihuahua could, with enough determination and a little help, successfully mate with a Russian wolfhound. That a Caucasian man and an Ethiopian woman, to take a random example, both being *Homo sapiens*, could meet, mate, and produce healthy, fertile offspring meant to Blumenbach that *Homo sapiens* was unquestionably "the most perfect of all domesticated species."

It's difficult to know what Blumenbach meant by "most perfect." Perhaps he thought that, just as there were degrees of humanness even though all races were human, so there were degrees of perfection among domesticated species. Or maybe he simply thought we were the most domesticated: cats and dogs are both domesticated, but dogs seem somehow more domesticated than cats.

Few writers since Blumenbach have considered the implications of human domestication, and of course we don't think of ourselves as a domesticated species. Darwin devotes the first chapter of *The Origin of Species* to "Variation Under Domestication" but, although we know he read Blumenbach, uses pigeons, not humans, as his example of domesticates. However, in the original paper on variation, which he read to the Linnean Society in June 1858, the year before the publication of *Origin*, he distinguished between species living in the wild and those living under domestication in a way that could hardly have failed to suggest to which category humankind belonged: "The two are so much opposed to each other in every circumstance of their existence," he wrote, "that what applies to the one is almost sure not to apply to the other. Domestic animals are abnormal, irregular, artificial; they are subject to varieties which never occur and never can occur in a state of nature; their very existence depends altogether on human care." In *Origin*, he suggested that if a domesticated species were somehow returned to a state of nature, it would either revert to its original wild form or else become extinct.

Samuel Butler wrote the Utopian novel *Erewhon* in 1872, thirteen years after *Origin* and one year after Darwin's second great work, *The Descent of Man*. (The title is an anagrammatical pun on the word *Utopia*, which is Greek for "no place"; *Erewhon* is "Nowhere" spelt backward, almost. Why he didn't just spell it *Erehwon* remains a mystery; if he did not want his fictional lost world to be seen as simply the real world read backward, then why did he give its citizens names such as Yram, Senoj, and Nosnibor?) Butler was an ardent Darwinist, and in his novel he applied Darwin's notion of domesticity to our own species. He posited that human beings have been domesticated by their own machines, what today we would call technology. In *Erewhon*, Butler's narrator, Higgs, stumbles across a society that has turned back the clock by having outlawed every mechanical advancement made in Europe during the previous 271 years, or since the beginning of the Scientific Revolution in 1600. They had spoons, shovels, and horse-drawn carts, but no trains, clocks, or steam engines. (Higgs is arrested and jailed for owning a pocket watch.) Curiously, they also had no religion. Moral issues were settled by philosophers or simply referred to custom. Butler believed that, outside Erewhon, machines had taken on a kind of independent life of their own. He even ascribed Darwinian classifications to them in terms of their species (knife, for example), subspecies (carving knife), varieties (serrated carving knife), and even races (Japanese serrated carving knife). And he ascribed to machines a degree of intelligence – "Who can say that the vapour machine [the steam engine] has not a kind of consciousness? Where does consciousness begin, and where end? Who can draw the line? Is not everything interwoven with everything?" – as well as the ability to reproduce themselves (using humans as midwives): "Do we not use a machine to make a new part for a machine?" According to Butler, humans, like domesticated animals, had become so dependent upon the artificial environment created for them by machines that they could no longer (again, outside Erewhon) survive without them. Machines had become man's "extra-corporeal limbs"; removing them was a kind of amputation. Echoing Darwin, he suggested that "if all machines were to be annihilated at one moment . . . and all knowledge of mechanical laws were taken from [us] so that [we] could make no more machines, and all machine-made food destroyed so that the race of man should be left as it were naked upon a desert island, we should become extinct in six weeks."

Unlike their European counterparts, Erewhonians – whom Butler located in central New Zealand – halted technological advance in its tracks before it had progressed so far as to turn humans into automatons. This did not make them savages, only slightly backward, like their names. They were in fact rather admirable. It's possible that Butler, who spent many years in New Zealand, knew about Japan's three-hundred-year rejection of technology. Like Erewhon, Japan was a closed society – very few foreigners were allowed in, and none were allowed to travel freely within the country. Guns had been introduced to Japan in 1543 by Portuguese traders, and at first were adopted with enthusiasm. Japanese gunsmiths quickly became the best in the world. By the early 1600s, however, all gunmakers were required to move to Nagahama and produce firearms only for the government, and then gradually the government stopped ordering them. Shoguns refused to use firearms, claiming that they demolished the difference between samurai and peasant, and thus upset the whole balance of Japanese feudal society. During the Shimabara Rebellion of 1637, when twenty thousand Christian converts with five hundred flintlocks took over Hara Castle, the insurgents were savagely put down by sword-bearing samurai, and that was the last time guns were used in Japan until Commodore Perry sailed into Tokyo Bay in 1853. As Noel Perrin writes in *Giving Up the Gun*, "men can choose to remember; they can also choose to forget." The Erewhonians chose to forget.

But it is no longer possible simply to forget our reliance on technology. As John Livingston points out in *Rogue Primate*, "human domestication is, nearly enough, a synonym for civilization," and we can only occasionally, and briefly, forget that we are civilized (usually with disastrous results). That was the burden of Freud's *Civilization and Its Discontents*: as individuals, we have brief, mental flashbacks to our pre-civilized selves, and they drive us insane. Freud defends civilization on the grounds that it has conferred great improvements on our physical well-being: longer, healthier lives, a steadier supply of food, adequate shelter from "the harsh effects of nature." In short, civilization offers the same benefits that animals receive when they are domesticated. But he also notes that these accomplishments have been achieved at a great cost to our psychological health, in that we are constantly called upon to confront, and alarmingly often fail to resolve, conflicts between our natural, individual desires and the communal behaviour imposed upon us by society. "The word 'civilization,'" he writes, "designates the sum total of those achievements and institutions that distinguish our life from that of our animal ancestors." It thus measures the difference between wild and domesticated. He cites technology as the prime example of our "cultural acquisitions," noting that, by extending our ability to control the external forces of nature, it has given us "god-like" stature. Echoing Butler, he notes that thanks to technology "man has become, so to speak, a god with artificial limbs." He predicts that "distant ages will bring new and probably unimaginable advances in this field of civilization and so enhance his god-like nature. But in the interest of our investigation let us also remember that modern man does not feel happy with his god-like nature." Domesticated humans still yearn for individual freedom (which for Freud meant unfettered sexuality), and that yearning is the source of our neuroses.

Most of Freud's patients came to him, he believed, because of the conflict between their own natural sexual selves and the restrictions imposed upon them by civilization. They wanted the freedom to mate at will with any preferred partners, including family members (as many wild primates do), but they were prevented from doing so by invented social mores. The resulting split made them psychotic. Generalizing this split, he writes that "much of mankind's struggle is taken up with the task of finding a suitable, that is to say a happy, accommodation between the claims of the individual and the mass claims of civilization." A successful accommodation, however, involves such high levels of sublimation and the suppression of our most powerful libidinal drives that the effort to achieve it causes huge psychic damage. We recognize the need for the leash, but we strain against it. Accommodating to civilization "cannot be done without risk," Freud writes. "One can expect serious disturbances."

There are many manifestations of these disturbances. In fact, once we think of human beings as a domesticated species, many things begin to make sense. Take our complicated relationship with the natural world, for example. Much as we might pine for the wilderness, as domesticates we don't really belong in it. As Ronald Wright puts it in *A Short History of*

Progress, "as we climbed the ladder of progress, we kicked out the rungs below. There is no going back without catastrophe." Domestic animals like ourselves live in highly controlled surroundings, protected from the vicissitudes of life in the wild. They lead safe, regulated, repetitive, predator-free, depathogenated lives. Most of their food comes to them in convenient packages, sterilized, nutrient-enhanced, and medicated. At my home in Eastern Ontario, the feed I give our chickens comes in 55 kilogram bags and contains the proper proportion of corn, barley, soymeal, and other ingredients we call "natural," to give them exactly 16 per cent protein, for good muscle growth, and enough calcium to ensure thick eggshells. No chicken would ever be able to find such convenient fodder in the northern wild. As Darwin and Butler noted, take a small group of domesticated animals out of their artificial environment and plunk them down in the middle of an undisturbed wilderness and they wouldn't last a week, perhaps not even a night (think of that tethered goat in *Jurassic Park*); certainly not long enough to reproduce and perpetuate their kind. And yet that is exactly what happened five centuries ago, when European settlers first came to North America. So why are we still here?

Darwin's choice – return to type or go extinct, go feral or die, the motifs of *Tarzan of the Apes* and *Lord of the Flies* – is incomplete: there's a third choice. When Europeans landed on the shores of North America, a domesticated species released from our walled, sanitized, and regulated European compounds, they were stranded on an unknown shore surrounded by nature redder in tooth and claw than anything Tennyson could have imagined. Some of them went feral and some of them died, but, being "the most perfect of the domesticated animals," the rest of them came up with an alternative: remain civilized by subjugating the chaos around them and turning it into a more suitable environment for domesticated animals and their domesticated selves. Build walls, rid the enclosure of predators and competitors, replace the unfamiliar with the familiar, exercise control. We cut down the trees and planted new trees, we tore up the grass and sowed new grass, we removed the animals and introduced new animals. Not just in North America, but in South America, Australia, New Zealand, anywhere where anything just grew, wild. Today, we are still in the process of removing all wild species in the world and replacing them with domesticates.

Wild animals have little control over their own environment. They might build dams or nests to protect their young from exposure to weather or predators, but just about anything that sets out to get at them will do so. I once watched a pair of chickadees flutter noisily but helplessly around their nest as a black rat snake, which I had previously watched slither straight down from the roof of our log cabin and up the trunk of the chickadees' tree, ate each of their three nestlings in turn. About the only way most wild species have of ensuring that enough of their genetic material survives into the next generation is to have lots and lots of offspring, which is why chickadees can lay a second clutch of eggs, many plants send off millions of seeds each year, fish deposit their ova by the thousands, and rodents have huge litters. In other words, they don't try to control their environment, they control themselves. They adapt.

We don't. We're pretty much the same people we were when we domesticated wheat and corn and goats ten thousand years ago, probably not much different from the people who made firepits in Brazil fifty thousand years ago. John Livingston identifies a number of pre-existing traits that make some animals ideal targets for domestication: placidity, fecundity, rapid growth, dependency. They must be herd or pack animals — essentially, they must recognize some hierarchy to which they are prepared to comply. Bears, cougars, moose, owls, are solitary creatures. They have regular habits and habitats that keep them focused on their own needs. Cattle, sheep, goats, camels, caribou, horses, wolves, live in social communities and thus domesticate easily. There are exceptions, of course: African elephants are as social as Indian elephants, but have never been successfully domesticated. In most cases, however, domestication is the process of replacing the animal's natural "leader" with a designated human: my neighbour is his dog's alpha adult. Which means that the dog is a beta, an adult in body but a perpetual adolescent in every other respect. Adolescents have a curious ability to be both fiercely individualistic and slavishly conformist at the same time. As market analysts know, adolescents want something that will make them stand out in a crowd without setting them apart from the crowd. Researchers who have raised coyote pups as pets report that adolescent coyotes behave exactly like domestic dogs – tail wagging, barking, face licking, subservience, exuberance, dependence – until they pass out of adolescence into adulthood, at which point they simply trot off into the bush and are not seen again (unless they come back for the chickens). The trick to domesticating animals is to arrest their mental development at the adolescent stage. "The domesticated mammal," Livingston writes, "... is docile, tractable, predictable and controllable." All adolescent traits.

The alpha for humans is an abstraction, an idea. Livingston notes that throughout history we have been in thrall to some idea, some "ism" or other: imperialism, nationalism, romanticism, corporatism. Socially, like many domesticated animals, we are a faddish species. We choose an arbitrary leader and follow it into the barn. To become what Malcolm Gladwell, in *The Tipping Point*, calls a "social epidemic," an idea doesn't have to be brilliant. High heels and neckties are not brilliant ideas. Pet rocks and Ookpiks were downright dumb ideas. But they played to the adolescent desire in all of us to conform, to be like everyone else only different. Beer manufacturers know this. A beer executive I spoke to once assured me that the last thing he wanted his company to do was come up with a beer that was better than the number-one-selling brand on the market. Heaven forfend that his beer should be qualitatively different from one that was already a proven success. All he wanted to do was to make his new brand distinctive in some abstract way: a differently shaped bottle, a snappier label, a cap that twisted off. A beer that belonged on the same shelf as all the rest, but with an indefinable something that made it stand out.

Ideas are extensions of ourselves, and technology is the transformation of an idea into tangible reality, into a thing. Echoing Butler and Freud, Livingston refers to technology as our "prosthesis," the artificial extension of our natural reach. We are, he says, a "prosthetic species." We began to become domesticated the minute we "tamed" fire: "The domesticator of our ancestors," Livingston writes, was the "intensification of their dependence on technical knowledge beyond anything previously experienced." We became seduced by our own ability to conquer our environment. When we gained control of fire, we were suddenly able to keep warm, and therefore could inhabit regions that before had been inhospitable; we could clear land for agricultural purposes; we could boil water to soften grains too hard to chew. With fire, we could change all sorts of environments to suit our needs, rather than change ourselves to suit different environments.

(Here the story of Prometheus is instructive. Prometheus was the Greek mortal who stole fire from the gods and gave it to humans. Given that this act made civilization possible, you'd think Prometheus would occupy a special place in our regard. You'd think we would have been grateful. Instead, the story has Prometheus punished almost beyond endurance by the angry gods: he is chained to a mountaintop, and every day two vultures descend upon him and tear out his perpetually regenerating liver. True, it's the gods, not us, who torture him, but no human tries to save him. Presumably, Prometheus's ordeal would have ended if we simply gave the gods back their fire, but nobody gave the gods back their fire. (And similarly, in Christian mythology, it was an angry God who punished the first humans by expelling them from the Garden of Eden, not for stealing fire but for tasting fruit from the tree of knowledge of good and evil, which are abstractions. There are no accounts in the Bible of Adam and Eve or their descendants trying to fight their way back into Eden. They didn't need to: they already had knowledge. They used it to plough the land and raise cattle. They used it as they used fire: to domesticate themselves. "Utter dependence is common to all domesticates," writes Livingston. "Non-human species depend wholly on us; we depend wholly on storable, retrievable, transmissible technique." Knowledge.)

The two big Ideas that have been constant with us at least since the beginning of civilization have been religion and science. It is no accident that both have as their ends the control of nature. For most of recorded history, one or the other has predominated; that is, we have taken our domesticating ideas either from religion, or from science. In ancient Greece, science dominated thinking: any time a man can spend most of his life working out the mathematical relationships among the strings of a lute and be called a philosopher, you know you're in an age of science. From the seventeenth century to now, science again held sway. Now, however, when the elected leader of the most powerful nation on Earth is asked who his favourite philosopher is, and he answers, "Jesus Christ," or when the bestselling work of fiction of the day is an incredibly badly written potboiler whose main tenet is that there may be direct descendants of Jesus Christ alive among us today, or when research into stem cells is curtailed by government edict because its reliance on tissue samples from aborted human embryos is deemed offensive to public morality, you can be fairly confident you're in an age of religion.

If we apply Freud's sources of individual neuroses to society as a whole – Social Freudianism, as it were – we can see the swings from ages of religion to ages of science as analogous to the individual's vacillations between the primitive urges of the libido and the more rational dictates of civilization. Religion as primal urge, science as civilizing impulse, the two sparring off against each other in our collective psyche and doing us, as a species, real harm. But in fact the conflict is rarely between religion and science: it is almost always between dogma and science. Dogma is the application of religious beliefs to controlling an environment. Believing in a Supreme Being is religion; insisting that that Supreme Being created the universe in 4004 BC is dogma. Dogma can never co-opt science, because the only products of pure science are knowledge and, more often, more questions. Dogma can, however, co-opt technology, which historically has been the implementation of knowledge gained through scientific inquiry. Technology can be implemented to further the ends of dogma, as when, say, the rack is used to loosen the tongues of heretics, or seeds are genetically modified to expand a country's economy into foreign markets. And technology, as we have seen, is the well-spring of ideas that control us as a domesticated species. Dogma and technology are a dangerous combination, and they are the two main forces now driving Western society.

Using fire to warm a cave or make a clearing isn't science;

it's technology. Turning base metals into gold is technology. Genetically altering a potato is technology. Technology used to be the practical application of knowledge acquired through science: the scientist in her lab coat and pocket protector would shout *Eureka!* on discovering a bug that eats the bug that causes dysentery, and the biotech boys would set about making a billion pills containing said bug to sell to dysentery sufferers around the world. That is no longer happening. With the aid of computers, which are themselves the products of technology, technology is now taking place not at the speed of science but at the speed of light. "Technology," says physicist Barry Commoner, commenting on the proliferation of genetically modified organisms (GMOs) in our supermarkets, "has now gone well beyond what science knows." In other words, whereas in the good old days science could mull a problem over for a month or a year or a decade until it was certain it had considered as many angles as the problem presented before handing the solution over to technology – could find out, for example, what else the bug kills besides the bug that causes dysentery – nowadays technology makes a thing first and leaves science playing catch-up, trying to figure out what the unforeseen implications of the thing are. And, as Butler saw, technology can now reproduce itself: you cannot make a new computer without using a computer. We don't even make new cars without computerized robots.

For Commoner, GMOs represent technology out of control, technology in the service of the dogma of progress at any cost. Biotech companies are churning out GMO soybeans, potatoes, wheat, corn, and canola, and no one, he says,

knows what those plants will produce three or forty generations down the line. A potato plant with a modified gene that
produces a protein that kills spiders may kill only spiders in its
first generation, but the gene could mutate in seed from that
plant and start producing proteins that kill anything. According to Commoner, the entire concept of genetic engineering
is founded on a false premise; that genes always produce the
specific protein they are coded for. They don't. In standard
gene theory, a gene's nucleotide sequence, or code, is transmitted to a distinctive amino acid sequence to create a particular protein. In many organisms, however, up to 40 per
cent of their genes are alternatively spliced, which means
"the gene's original nucleotide sequence is split into fragments that are then recombined in different ways to encode
a multiplicity of proteins, each of them different in their
amino acid sequence." A gene from a benign bacterium, such
as *Bacillus thuringiensis*, which is now being transplanted into
potatoes to kill spiders, might at some future time, in response
to who knows what subparticular trigger, "give rise to multiple variants of the intended protein – or even to proteins
bearing little structural relationship to the original one, with
unpredictable effects on ecosystems and human health." The
result, he says, "could be catastrophic." This may well be the
real reason that biotech firms like Monsanto are so desperately trying to prevent farmers from collecting seed from
their GMO crops: not because the farmer would be violating
Monsanto's patent, but because seeds from genetically modified plants may produce plants that are toxic to humans.
Imagine the lawsuits.

Cloning a sheep or a human being is also technology. Science tells us that every individual cell in the human body contains all the genetic information needed to make that body. Technology says, So, if I take the nucleus out of a cell from, say, Arnold Schwarzenegger, and put it in a cell taken from a human egg, put the renucleated cell back in the egg, and put the egg in a human womb, the child born from that egg will be an exact duplicate of Arnold Schwarzenegger? Science says, Yes, but why would you want to do that? But technology is already in the next room looking up Schwarzenegger's phone number. The problem, as a growing number of scientists are beginning to perceive, is that there are a lot of disquieting answers to that troublesome Why question. An aesthetic answer: someone in Hollywood, for example, might think Arnold Schwarzenegger had the kind of body and mind that should be preserved in perpetuity. A militarist answer: a thousand Arnold Schwarzeneggers might make a formidable assault team. A political answer: because then we'd have an Arnold Schwarzenegger who was born in the United States.

Cloning is the ultimate control of nature. When Robert Briggs and Thomas King cloned the first frog, they took the original nucleus from a frog embryo, which means that the donor could not have been chosen for any particularly desirable traits other than its irreducible frogness. But if the original nucleus had come from an adult, that adult could have been chosen for its superlative frog qualities. It could have been the Arnold Schwarzenegger of frogs. Briggs and King might have disliked the sound of croaking, for example, and chosen a mute frog as a donor; they could have produced an entire pond of mute frogs. (Think of my neighbour's dog.) They could have chosen a member of an endangered species of frog, of which there are an ever-increasing number. Or, if they had culinary inclinations, they might have selected a frog with extraordinarily large, meaty legs. Or a frog with the right skin colour for attracting pickerel or great blue herons. Who would have objected? Certainly not researchers at Texas A&M University, who in May 2003 cloned an adult white-tailed deer, a perfect male specimen of *Odocoileus virginianus* named Dewey, whose nucleus-donor had been shot on a Texas game farm where hunters pay up to twenty thousand dollars to kill exceptionally well-endowed (antlerwise) white-tailed deer. When Dewey reaches sexual maturity, he'll be put to work as a stud deer, as are prize bulls and winning race-horses, his semen sold to owners of other game farms who will use it to stock their farms with Dewey offspring. This is the instant domestication of white-tailed deer: it took ten thousand years to make the perfect Holstein bull; it took 212 days to make Dewey.

The fear is that cloned human beings will be more like Dewey than like the twenty-seven innocuous tadpoles produced by Briggs and King; more like Arnold Schwarzenegger than like, say, Michael Moore. It would be possible to clone an entire army of unhesitant, unscrupled soldiers, for example. Novelist Ira Levin, in his novel *The Boys from Brazil*, imagined just such a scenario as early as 1976 – an entire army of cloned Hitlers. Corporate-sponsored geneticists could clone a whole town full of contented Wal-Mart shoppers: no need for clumsy robots, as in *The Stepford Wives*, and no worries about escapees, as in *The Island* (in which cloned duplicates of "real" people are kept in a special sanctuary in case their nucleus-donors ever need new body parts).

Human cloning may be much closer than the horizon. Several companies have already been set up to take orders (and tissue samples) from clients who want to perpetuate themselves or their favourite aunt the minute the technology becomes legal, much as other companies are compiling passenger lists for the first commercial flight to the moon. Walt Disney needn't have had his carcass frozen if cloning had been an option. Some investigators, such as Elaine Dewar, author of *The Second Tree*, an unquiet assessment of where we are on the path to human cloning, are convinced that a human Dolly the Sheep already exists somewhere or, like Dolly, existed for a brief time and then died. In November 2001, a U.S. company, Advanced Cell Technology (ACT), announced that it had managed a successful human-cloning experiment: its technicians had taken a nucleus from a client's skin cell, injected it into an unfertilized human egg from which the nucleus had been extracted, then got that egg to divide in the usual fashion. If the egg had gone on dividing, it would have eventually produced a human baby. As it was, the egg died after the second cell division, producing nothing but a new definition of the adjective "successful."

Two thousand and one was a big year for cloning. In August, the American National Academy of Sciences sponsored a panel on cloning in Washington, D.C., and the Canadian government created a Standing Committee on

Health, which held hearings on a proposed law that would ban the cloning of human beings and restrict stem-cell research to therapeutic rather than reproductive cloning: in other words, research leading to stem-cell cures for diseases such as Parkinson's could proceed, but research leading to Dolly the Human Sheep could not. The following year, in December 2002, the Raelians – a UFOlogical cult based in Quebec that believes human beings were cloned from an alien species twenty-five thousand years ago (which, if true, would make cloning very old hat) – announced that they had brought a cloned human baby, which they named Baby Eve, to term in a surrogate mother. But since no corroborative evidence materialized, no Baby Eve smiling into the cameras, their claim is now thought to have been a hoax. But the interesting thing about the Raelian episode is that they had apparently found fifty women who were willing to act as surrogate mothers for nucleus-transferred eggs, presumably in order to produce fifty copies of their leader, Rael, a former French racing-car driver named Claude Vorilhon who saw the light when he was abducted by aliens, and who believes that religion and technology will control the future. The fifty willing women were offering their bodies for religious, not scientific (or technological), reasons.

A growing number of people believe that human cloning is inevitable, and that any ethical objections to it are either ill-considered or beside the point. Thus Philip M. Boffey, a senior science writer with the *New York Times*, writes that no one will ever be able to clone an army because doing so "would require a huge number of women to supply the eggs and bear the fetal clones to term." The Raelians seem to have overcome that obstacle before Boffey even thought of it. Boffey also asserts that a clone of Hitler, raised under different circumstances from the original, "would not have the same career trajectory." The clone would have a different personality. Based on what we know of monozygotic twins separated at birth, however, we cannot assume this. Identical twins reunited after thirty or forty years, people who didn't even know they were identical twins, often find that they have lived astonishingly parallel lives: they have the same jobs, drive the same cars, marry spouses with the same name, own identical pets, wear similar clothing, have the same food preferences. Who is to say they don't also have the same dreams?

The only real argument against cloning, Boffey implies, is that it presents a safety risk, since so many animal clones have been born with serious health defects. Boffey dismisses this objection. Once a "renegade" group like the Raelians or ACT, or technicians working in South Korea or Singapore or the United Kingdom (where reproductive cloning is not banned) presents us with a healthy, happy child with blond hair, blue eyes, and perfect teeth, he says, "the safety argument becomes less persuasive." But the safety argument as well as some strong ethical objections have already been compromised. At least one fertility doctor, Alison Murdoch of the University of Newcastle upon Tyne, has received permission to clone human embryos from persons with certain genetic diseases so she can watch how their stem cells develop, and then test various drugs on them. Viewed in such a light, testing drugs on human clones could seem safer in the long run than testing them on animals. But what a glaring light it is: you don't have to be a former French racing-car driver to envision a future in which drug-testing facilities at universities and pharmaceutical companies have human clones rather than chimpanzees and orangutans in their holding cells.

The fear that cloning would be used by unscrupulous people for personal ends is a real one, given human history. Because cloning offers such complete control over our environment – as Bill McKibben suggests, we can stock it with people whose personalities, body types, consumer preferences, voting tendencies, even shoe sizes, are predetermined – and because it is a product of the same technology of which we are a product, if the only thing hindering human cloning is the ethical compunction of a few scientists or Utopian novelists living in an age of religion, it will happen. Isaac Asimov, in his novel *I, Robot*, foresaw a time when thinking robots worked on automobile assembly lines, cotton plantations, hazardous waste recycling stations, and other places where boredom or rebelliousness or danger make it difficult to find and keep skilful and alert workers. It is now more likely that such jobs will be done by clones. Asimov's robots rebelled; Walter Mitty clones wouldn't. It's not as though we've balked at the ethics of it in the past: we've already classified the Negroid race as subhuman in order to clear our consciences for slavery, and we still use Asian children to knot rugs, pick fruit, and separate heavy metals out of recycled batteries and computers. How much more humanitarian it would be to get clones to do it. Clones could easily become morally acceptable as second-class citizens. We could deny them the right to vote, hold office, have children the "normal" way. They would be a class of drones, kept around for drudge work and spare body parts, but segregated from real people. We would allow clones to work in our mines or in our factories, but we wouldn't want our daughters marrying one of them.

No crack is too small if the edge of the wedge is thin enough.

In 1902, thirty years after the publication of *Erewhon*, Samuel Butler wrote *Erewhon Revisited*, in which Higgs returns to Erewhon to find that in the years since his first visit, the Shangri-la innocence of the isolated valley has been completely turned on its head. The first Erewhonian he meets is dressed in the English fashion, pulls out a pocket watch to check the time, and talks about the forthcoming dedication of a great temple. Higgs is distraught to learn not only that the laws against machines have been repealed in the new order, but also that a fundamentalist religious movement has sprung up, the idolized god of which is Higgs himself; a mythical deity who visited Erewhon twenty years previously and miraculously ascended into Heaven (Higgs had escaped from Erewhon by making himself a hot-air balloon). The two new forces – mechanism and Sunchildism, as the new religion is called, the high priests of which are the managers of the Musical Bank – are so fused in the imagination of the people that the one is expressed in terms of the other. In a treatise on free will, Erewhon's president proposes that "if, for example, A's will-power has got such hold of B's as to be able, through B, to work B's mechanism, what seems to have been B's action will in reality have been more A's . . ." Death and illness are characterized as failures of technology: "A's individual will-power must be held to cease when the tools it works with are destroyed or out of gear. . . ." The Erewhonians are docile, tractable, predictable, and controllable, and hence vulnerable to totalitarian rule: in the new Erewhon, greed is rampant, corruption of power everywhere evident, suppression of truth the order of the day. It is all chillingly familiar. When Higgs identifies himself as the returned Sunchild and tries to point out the degree to which his actions and words have been misconstrued and manipulated for certain individuals' personal gain, he is imprisoned and sentenced to death. He escapes, returns to England, tells his tale to his son, after which, deemed insane, he dies.

Noel Perrin writes that Japan's "two hundred and fifty years of technological retrogression may seem to have no great significance, except as a historical curiosity," although it does offer "proof that a deliberate turning back is in fact possible in a civilized society." But how can that be possible in a society so civilized it no longer knows who its oppressors are? We are not like the domesticated animals in George Orwell's *Animal Farm*. When our oppressor is our own domesticity, against whom do we rebel?

Ronald Wright suggests that as past civilizations began to decline, instead of cutting back, taking stock, changing their paths to total collapse, "they dug in their heels and carried on doing what they had always done, only more so." The Sumerians destroyed the Fertile Crescent, the Easter Islanders built ever bigger and more resource-draining statues, the Maya turned to "higher pyramids, more power to the kings, harder work for the masses, more foreign wars." Why? Why not say, to use Bill McKibben's powerful word, "Enough!" Wright's answer is a bit like Freud's: because civilization works. Despite multiple collapses, there are thirty times more of us on Earth now than there were during Roman times. Like Prometheus's liver, nature is regenerative, seemingly inexhaustibly so. We can just move somewhere else and wait for the soil to become fertile again, for the forests to grow back, for the population cycles to reboom.

But there is a more depressing answer. Civilizations may have carried on doing what they had always done simply because that was all they knew how to do. When sheep in their enclosure have eaten every blade of grass but one, they don't try to come up with some other way of surviving: they eat that last blade. Darwin's choice of going back to some original state of nature is really no choice at all. As the most perfect of all domesticated species, we can never go back to being what we were before. Our only option is to go on, to domesticate everything around us as quickly as possible. Once the entire world is domesticated, we'll be all right. If things begin to fall apart before we get there, if the centre doesn't seem to be holding, step up the pace. Like habitual gamblers down to our last pile of chips, we don't cash them in and go home; we go all in.

My neighbour's dog is quieter now. It seems to have barked itself out. At night, when the coyotes commence their yipping, it sits on its haunches in its compound and whines. Far from being terrified of coyotes, the dog now wishes only to join them. It seems to have realized that any attention it's going to get will come from them, not from its owner. My neighbour is glad about this, because he is training the dog to hunt coyotes. One day he will place a radio collar around the dog's neck and let it go. The dog, following its libido, will lead my neighbour to the coyotes, and my neighbour, following the dog, will kill them.

SAFE AT HOME

After a week of living in a tent, each night pitched in a new and seemingly wetter and windier location, it feels good to be sitting in calm sunshine in a baseball park on a Friday afternoon, my binoculars around my neck, a Women's Auxiliary hot dog in my hand, and a cup of Tim Hortons coffee balanced on my knee. I've been car-camping my way to Newfoundland, taking my time, exploring some of the lesser travelled routes, watching a few birds, slowly defragging myself from the somewhat jumbled familial entanglements that we call being home. Last night, I camped in a remote park on the Quebec/Maine border and watched a thunderstorm fill the entire sky above my head with a dark, blue-and-yellow bruise. In the morning, I squeezed out my gear and drove south to New Brunswick, and now I'm in Moncton, celebrating the improved weather by taking in a Triple-A baseball game.

My mother died a few weeks ago, of cancer. She was born in Newfoundland, and although she lived in Ontario for the last fifty-eight years of her life she never stopped referring to Newfoundland as "home." I'm on my way to Newfoundland on a kind of pilgrimage, as though being in the place she called home will restore a part of her to me, perhaps even show me a part of her I hadn't known.

To me there are few better places in which to meditate on the meaning of an abstract concept than in a ballpark on a sunny afternoon watching two teams engage in the sort of choreographed chaos that is baseball. Moncton's Kiwanis Park is the home of the Moncton Cubs. I'm not going to delve into an extended disquisition on baseball as a metaphor for life, but as a balletic still point in the eye of the fractal storm, baseball has few rivals. Even Triple-A baseball is applied poetry. Time is not a problem in baseball. It is one of the few sports, as baseball writer Roger Angell has famously observed, that is ungoverned by a clock, which is what most of us long to be. Unless the field is wet enough to attract the interest of Ducks Unlimited, each game is played until someone wins, even if it takes all night and most of the next day, even if a team runs out of pitchers and has to send in the bat boy. Both baseball and life recognize that a proper ending is one for which time should not be the deciding factor.

Baseball is also the only sport I can think of in which the much-touted home-team advantage really is an advantage. In baseball, this has less to do with home-team players luxuriating in the psychic flow from a choir of sympathetic fans (there were only twenty-three of us in the stands, a number so low as to be psychologically damaging rather than the opposite) than it does with the fact that the very rules of the game are so structured as to give the team playing at home an edge over the team playing away. If you are at home and are losing, you always have an extra half-inning in which to come back and win, whereas if you're away, you don't. My mother would have appreciated that. Only the away team experiences sudden death.

For many animals – migrant birds, for example – home is neither an abstract concept nor even a roughly defined local territory. It is about as precise a spot as it is possible to imagine. GPS has nothing on some birds. Summer after summer, generation after generation, the Eastern phoebes (*Sayornis phoebe*) at our house in Eastern Ontario return from southern Mexico to the same nest, constructed on a downspout where our kitchen joins the main house, even though I've made them a perfectly adequate nesting platform over which they must fly to get to their louse-infested clump of woven grass and mud. For those phoebes, home is where that nest is. In a study of Thick-billed murres (*Uria lomvia*) in the Canadian Arctic, researchers banded fourteen birds on Prince Leopold Island and checked the following year to see if the same birds returned to the same island after spending the winter off the east coast of Newfoundland, where my mother's brothers would have called them "turrs" or "baccalieu birds." The study found that not only did they return to Prince Leopold Island, but each nesting pair returned to within five centimetres of the nesting site they had occupied the year before. Getting the old site back was so important to them that, in one case, when a pair was forced by intrusive newcomers to nest three whole metres from its old site, the female was so traumatized she did not breed.

Few of us, fortunately, are that site-specific when it comes to defining what we call home. We can and do breed just about anywhere and at any time – a characteristic that confirms our status as a domesticated species. In other words, it may be easier to domesticate a species that doesn't really care much where it lives. Chickens and pigeons, for example, unlike murres, will lay eggs anywhere. Socially, humans have adopted a structure peculiarly close to that of canines, a savannah animal whose social arrangements we may have imitated when we graduated from the subtropical forest. Canines have a much looser idea of home than do, say, baccalieu birds.

The coyotes (*Canis latrans*) that live across the road from us, for example, have a three-tiered concept of territoriality that seems complicated but works well and is actually fairly low-maintenance. They occupy a large home range, the size of which is not strictly defined but depends on such variables as food abundance, pack size, and terrain. The perimeter of this large range is scent-marked and regularly patrolled, but not rigorously defended; when another coyote enters on its way to somewhere else, its presence is duly noted, mildly protested, but the intruder is allowed to pass through without untoward incident – much as when a foreign vessel, say an

American icebreaker, sails into Canadian waters in the Arctic without officially asking permission to do so. A few politicians become mildly territorial about it, but no one calls out the heavy artillery.

Within that larger range is a smaller territory that is more actively occupied and defended by the coyote pack. This area will often have good hunting sites, as well as water and some nice southern exposures. It's where the pack sleeps and hangs out during the day (coyotes are crepuscular hunters) in the non-breeding season, and where from December to February its courtship and mating activities take place. Scent-markings here will be stronger and closer together, trails into it will be posted with scat, and any stray coyote happening along will know better than to venture into its precincts. (If this is beginning to sound like a Jack London novel, I apologize. I am not anthropomorphizing coyotes. Rather the other way around. London was good at describing certain affinities between canine and human behavioral patterns; where he erred was in suggesting that canines behaved badly, and that humans behaved badly when they behaved like canines.)

Somewhere in this smaller territory is the coyote pack's core area, where the den-sites are located. From March until June or July, the alpha female retreats there to have and rear her pups. Sentries are posted at the access points and intruders are strenuously rebuffed. Think of Sam Spade trying to talk his way into the backroom of the wrong speakeasy. In Banff National Park early one summer I was walking up the Bow River with a wildlife biologist when we came upon a lone coyote sitting calmly on its haunches, observing our approach. We thought it would move off when we got nearer, but it didn't. It just sat there. We stopped. It yawned. We stepped forward. It uttered a short yip. We stopped again, realized that our challenger was a first-year male on sentry duty, and made a careful detour around his guard-post without entering the area he was defending. I don't know what would have happened had we kept going; probably the admonitory yip would have become a warning bark, neck-hair would have bristled (his as well as mine), teeth would have been bared, and we would have been forced to beat a less than dignified retreat.

The core area contains more than one denning site. There may be five or six good dens within it, holes dug into hillsides, under boulders or the root-balls of fallen trees, or in the pick-up-sticks-like maze of wind-fallen trunks. The female and her pups occupy one den at a time, but not the same den for an entire denning season. She'll move the pups within the core area for reasons known only to her, perhaps on a whim or a premonition of danger. After our approach and the beta's warning yip, somewhere within the core area the pack's alpha female may have become jumpy and moved her pups to a den further from that particular access point. I'm guessing here. Den parasites might play a role, or an owl might begin haunting that particular neck of the woods. The point is, within the pack's core area there are a number of possible dens that are occupied on and off as long as the pack maintains its undisputed territorial claim. I suspect that if a coyote could be asked for its definition of "home," its answer would involve all three territorial types – the larger range, the inner region, and the core area – but it would not feel nostalgic for any one specific den-site.

All her life my mother felt nostalgic about Renews, her home port on Newfoundland's South Shore. As I sat in Kiwanis Park eating a ball-diamond hot dog and cheering for the Moncton Cubs, I wondered about those Thick-billed murres. Do they, as they make their way up the Labrador coast from their summer feeding waters, carry in their minds an image of the narrow ledge of rock they fully intend to occupy when they land on Prince Leopold Island or wherever it is they call home? It's hard to imagine they do, but the more I thought about it the harder it was to believe they do not. After all, my mother did for decades. How else could the murres return to the exact spot year after year? And why do they bother? Does it have something to do with being a migratory species? Could it be that the farther we stray from home, the more important the idea – our mental picture – of home becomes to us?

Early immigrants to Canada continued to write letters "home" to their countries of origin long after they'd been here long enough that they might be forgiven for thinking of Canada as home. I recently spoke to a man who came to Canada fifty-five years ago, from Glasgow, Scotland, who had been living in the same house in the same city since the day of his arrival, and when I asked him where he'd gone on his last vacation, he said, "I went home." My mother, who left Newfoundland in 1945, and whose brothers and sister I was on my way to visit when I stopped in Moncton to watch the ball game, always spoke of Newfoundland as home, never more so than during the last three months of her illness. "At home we always had our tea at four o'clock," she would say when I brought her tea. She still drank King Cole tea, and her sister sent her a package of hard tack every Christmas so she could make fish-and-brewis. She never actually made fish-and-brewis, but she was always threatening to, and she had a cupboard full of hard tack to give the threat some weight. When watching the Weather Channel, she would check the forecast for the Avalon Peninsula, where it was, invariably, "foggy along the coast with a 40-per-cent chance of precipitation."

At first this may seem to suggest that humans are more like murres than coyotes, that they have a very specific place in mind when they speak of home: a certain village, a fondly remembered city, a particular house. But often the very places we call home do not in fact exist outside our memories of them. There are immigrants to Canada whose villages, or even countries, of origin are no longer on the map. Glasgow today bears almost no physical resemblance to Glasgow in 1951, as my friend in Northern Ontario would be the first to tell you. As for my mother's home in Newfoundland, the image of it she carried with her was, like baseball, timeless; the outport she grew up in is all but deserted, and the house belonging to my grandparents – the one in which tea was always served at four o'clock, in which there was a fireplace in every room, in which my grandfather shaved tobacco from a plug with his penknife and tamped it into his pipe after dinner, and in which the parlour was always referred to as

"Jimmy's Wake," because a relative had in the dark backward and abysm of time been laid out in it when he died – has long since been torn down, its contents distributed, its cellar filled in. What, then, did she mean last year when she said, "I got a Christmas card from home"?

Well, she meant of course that she'd received a Christmas card from her brother, who still lives in Renews – in a new house, albeit one made partly with wood salvaged from the old one. For most of us, home, to paraphrase Jean-Paul Sartre, is other people. A young friend informs me that to her, home is where her parents live; her father is in the armed forces and her parents have moved every four years for as long as she can remember. Home to her is wherever they happen to be living at the time, even if she's never seen it. Similarly, a geologist I know, who spends nearly half of every year on remote rock outcrops in places like Australia and Namibia, says that home to him is three places: the Niagara Peninsula, where he grew up (but was not born); wherever he happens to be at the time, a tent on Beechy Island or a motel room in Adelaide; and wherever his wife and children are.

Our definition of home depends to a large extent on where we are when we are asked to give it. When I'm in Europe and asked where home is, I answer, "Canada." I am not being evasive; it is true that after a long trip abroad I experience a distinct sense of homeness when I push my luggage cart through the Arrivals lounge in a Canadian airport, even though I may still be a thousand kilometres from where I live. In canine-mimic terms, Canada is my home range.

If I'm already in Canada when asked about home, I'd have a different answer. Within my home range is a smaller territory that I occupy more actively. I won't call it a province, because it isn't defined by political boundaries, and anyway a province is composed of many parts, not all of which make me feel equally at home; my mother didn't say she was from "Newfoundland and Labrador." Home is more like a bioregion. Isabel Huggan, in *Belonging*, writes of "this carnal knowledge of landscape" that to her signifies home. I feel that visceral pull more strongly in a Great Lakes/St. Lawrence broadleaf forest than I do in, say, a British Columbian rain forest, or out on the shortgrass prairie, or even in a stand of Northern Ontario's balsam fir and black spruce. I understand the Laurentian forest. I recognize its plants, I know what birds and animals to look and listen for, what is safe to eat and drink, what to expect from its clearings and shadows. Being in a Laurentian forest takes an enormous amount of stress off the parts of my brain responsible for sensory data interpretation and survival responses. I may experience a sense of relief when I re-enter Canada at the Calgary airport, but I don't truly relax until I'm back in my own bioregion.

Within that bioregion, however, there is a core area in which I live. It isn't a single place, nothing as site-specific as a house or a village or a street. It's a triangle, the three angles of which are Ottawa, Montreal, and Toronto. I've moved about a dozen times within this area (I now live near Kingston, at its exact centre). I used to think my choice of den-sites was arbitrary, circumstantial – not really my choice at all, but responses to outside stimuli – but considered in terms of canine-mimicry it may be something deeper, and have somewhat profounder implications. It may explain, for example, why people who move to a place from some other place are often more alert to their new environment than are those who were born there. Like converts to a new religion, they demand more of it, are more consciously attuned to its nuances, less tolerant of its shortcomings because unexpected slippages pose the most threat to their own survival. And they may continue for decades to refer to their previous, safe environment as home.

The tiny outport of Renews rests at the tip of Renews Harbour, on Newfoundland's Avalon Peninsula, a cluster of brightly painted houses surrounding a narrow strip of water that for three hundred years has provided safe harbour for their seafaring owners. English colonists settled there in 1629. The Dutch destroyed the English plantation at Ferryland, a few kilometres up the coast, in 1677. By then, Renews was already well known to the transient Portuguese by the name *Ronhoso*, meaning "scabby"; when Jacques Cartier sailed past it in 1636, he recorded the French version of the name in his diary: *Rougneuse*. The name is thought to refer to the shell-encrusted rocks that still mark the harbour entrance at Renews Head, where wheeling herring gulls drop clams on the rocks to open them, as they have done since the beginning of time.

This was my third visit to Renews. The first was around my eighth birthday, when my mother brought me home to meet her family. On my second visit, forty years later, I had

had to stop at the store by the highway that now connects all the South Shore outports like a string of Christmas lights, and ask where the Goodridges lived. The storekeeper's first reply was, "Why do you want to know?" I explained that David and Wallace Goodridge were my uncles, and that I hadn't been to Renews since I was eight and couldn't remember how to find their houses.

"Oh," he said, "then you'd be Zoë's son." And when I said I was, he pointed across the harbour. "That's the Goodridges over there, the five houses under the trees."

I was wrong about not remembering. Just as there is more to a home than boards and stone, so history is more than a simple accumulation of time. As soon as I rounded the harbour the roads became familiar. Outport houses were built first on some convenient spot, and then the roads were made around them; somehow I knew to keep veering harbourward until I got to the old wharf, and then to turn uphill from there to the big house where my grandparents lived. Except my grandparents had died long ago, of course, and the big house was gone.

In 1820, Henry and Susanna Goodridge and their sons, Alan and John, moved to Newfoundland from the town of Paignton, Devonshire, and within ten years had established a trading company that quickly became one of the largest in British North America. By the end of the nineteenth century Alan Goodridge & Sons occupied an entire waterfront block in St. John's (although the family maintained their residence in Ferryland before moving to Renews), owned four merchant vessels outright and shares in four hundred others, and

every year bought and sold 120,000 quintals of cod – more
than 13 million pounds – and produced thousands of gallons
of cod-liver oil. By 1913 the firm employed three hundred
people, forty of them coopers. Under the family flag,
Goodridge ships sent cod and oil to England and brought back
tea and china, stopping in the West Indies for rum and sugar
and tobacco. My great-grandfather was the prime minister of
Newfoundland in 1896; his grand-nephew, my cousin Noel
Goodridge, was Chief Justice. The Goodridges are a proud
family, and their roots in Renews are as deep as the bedrock
that underlies the thin, rich soil and the harbour itself. There
were glaciers, then there were Beothuk, then there was
Renews, and, very shortly thereafter, there were Goodridges.

Uncle David was in his seventies. He built the house he
and Aunt Terry live in and in which they raised their sons, all
of whom now live in Renews within hailing distance of their
parents' house. There is no confusion in their minds about the
location of home. The five houses are their denning sites. In
the evenings Uncle David and I strolled through the village,
looking at the changes. On my first visit I could walk from
the big house to my grandfather's store at the head of the old
wharf for a bag of penny candy, my reward for splitting the
day's kindling. Once, instead of candy, I asked for a hook and
some line and went down to the wharf to jig sculpen between
the planks, until I accidently caught one and couldn't haul it
up through the crack. After that I wheedled Uncle David into
taking me out in his dory to jig for cod, and when we came
back I helped unload the catch into the splitting shed with
a pitchfork. Kettles of fish. Fish as big as I was. There were

stages on the shore beside the wharf, and cod flakes spread out on slats to dry, old gill nets poled over them to keep the gulls off the fish.

All gone, even by the time of my second visit. The old wharf, the big house, the store, the cod. The gulls were still there. Farther out over the harbour something bigger – a gannet.

Uncle Wallace never married. He stayed on in the big house looking after his mother, my grandmother, and when she died he and his brother tore the house down, board by board, in their silent, methodical way. It was taking up space. To them the loss of the house represented no loss of home. My mother was the same. With the wood, Uncle Wallace built his own house fifty feet from his brother's, living with them until it was finished.

We walked to where the old house had been. I could still see it under a clump of willows, a large, tall, white house overlooking the wharf. I remembered lying on the grass beside it, looking up at the sky and the clouds sailing above the eaves, and feeling pressed into the earth, as though the house, not the clouds, were moving. Now the lot where the house had stood was completely grown over. The company store was gone, too: "That's the hole where the cellar was," said David, pointing into an empty square in the ground, also mostly filled in with vegetation. "We boiled cod livers down there. That post rising out of the centre used to hold up the floor; it's the old mainmast from one of the schooners." All the old houses were built by boat builders: they laid the planking athwart the crosstrees. The old house was a three-decker, a Victorian mansion in a village of fishermen's houses. Uncle David stayed out of the family business, took up fishing even though he got seasick every day he went out. When he built his own house up the hill, he surrounded it with trees, fast-growing poplars, through which there is no view of the harbour.

We walked up to the top of the headland, called the Mount, for the dizzying view. Far below, a gentle surf scrabbled over boulders and turned to foam on a pebble beach. Another meaning of *rougneuse* is simply "rough." Uncle David pointed out into the harbour at a rock that just broke the surface at low tide. "A ship hit that rock once," he said, "with a load of bricks from England. Sank before anyone could salvage them." That was in 1872. He pointed to a pile of bleached wood beside the wharf. "One night a ship drifted into the harbour, not a crew member aboard of her, like a ghost ship, she was. They took her apart and used the timbers to build the wharf, and that's the rest of her there in that pile." That was in the 1890s. On the ground beside us, the round nose of a cannon poked out of the short grass like a turtle's head; black, pocked by salt wind, worn to a dull gloss. "Someone came down from the university a while ago and wanted to dig them up," said Uncle David, "but nothing ever came of it." There were four half-buried cannons on the headland, each pointing at a different approach to the harbour. "They say they were put here to defend against the Dutch." That would have been before 1677.

The next day we visited Uncle Wallace. He didn't walk with us, having become something of a recluse. Every morning, winter and summer, for years he got up at sunrise and walked inland to the woodlot, where much of the iron and timber left over from the old house was stacked — bedsteads, mantelpieces, fire grates, Canada thistles growing through the holes — and cut firewood until breakfast. Then he went home and was not seen again until the next morning. He read books, he told me. He kept a fire in the summer kitchen even in July. His house was a museum of local history. Books lined the walls of his living room. Along with photographs of Alan and John, paintings of two of the Goodridge schooners, the *Clementine* and the *Mayflower*, the Goodridge pennant flying from their topgallants, occupied a dining-room wall above a silver tea service set out on a side table, all salvaged from the wreckage of the old house. If Uncle David had an all-inclusive sense of time, time as an aura that surrounded a place, to Uncle Wallace time was an ocean in which he discerned the strong current of history. He had the original deed to the land on which the old house was built. He had Queen Victoria's letter appointing his grandfather Leader of Her Majesty's Loyal Opposition. He had his father's school scribblers and account books from the store. He had letters my mother wrote to her mother shortly after marrying my father and moving to the mainland.

The vanished house seemed to me to be a symbol of all that had been lost — in their lives, in my mother's, in my own. How, I asked him, could he have torn it down?

"I hated to see it deteriorate," he said, "but we couldn't afford to keep it up. While I was in England, vandals broke all the windows. Bats got in. It was just taking up space."

"You went to England?" I said.

Leaning forward in his chair, his elbows on his knees and his hands clasped as if in prayer, he said, "I thought I would travel. I thought travel would broaden my mind, but it didn't. I went to England to see our relations in Paignton. It seems they all became tavern keepers after our side of the family left." This would have nettled Uncle Wallace, who was a life-long teetotaller. "I was in London for two weeks," he said. "I walked every day. One day I walked into Foyles on Charing Cross Road and bought some books. Another day I took the Underground and got out somewhere near Highgate." He shook his head. "I saw a lot of poverty. The way people lived. I couldn't . . ." He looked out his kitchen window, toward his woodpile and beyond to his brother's house, to the fields that used to run with sheep, to where the remnants of the old house lay mouldering in the bunchgrass. "I just came back. In the end, you know, there really is no place like home."

Now, after my mother's death, on my third visit to Renews, I learn that Uncle Wallace has taken an apartment in St. John's for no reason that anyone can discern. He just announced one day that he was moving. He stays in St. John's during the week and drives back to his house in Renews for the week-ends. No one seems to have asked him why, and when I go over to see him on Saturday, I don't ask either. Instead, I wonder what he does in St. John's. "I read, mostly," he says. "And walk."

It's entirely possible that the ocean of time in Renews has become too much for him. For the past year, Steven Mills, a Memorial University archaeologist, has been excavating an old building site by the harbour, at the foot of the hill that once ran down from the back of the Goodridge house to the barn, and was therefore on Goodridge land. Mills and his field assistants — local high-school students working for summer wages — are exhuming bits and pieces of the Goodridge past, and Steven is generous and understanding enough to let me help. At first, he says as we dig, Uncle Wallace came down every day to see what they were turning up: parts of toys he may have remembered playing with, broken crockery from a long-vanished summer kitchen. As long as they were in the upper layers, Uncle Wallace kept up his interest. He brought Steven early photographs of Renews and favoured him with his vast knowledge of family history. Deeper down, however, below the sod line, when the researchers began finding items from the seventeenth and eighteenth centuries, mostly broken wine bottles, clay pipe stems, stemmed glassware, soot-covered stones from an ancient hearth, Uncle Wallace's interest declined. The building had evidently served as a tavern for many years. The Goodridges must have been tavern keepers in the New World as well as in the Old. This would have been an anomaly in Uncle Wallace's ocean of time, like a rain delay or a player's strike. The items, carefully collected, cleaned, numbered, and stored in the local schoolhouse, are on display for public viewing. My mother would have sided with her brother. She was always an intensely private person, not wanting anyone to know her business, as she put it. What Steven and his bunch are doing she would have called snooping, and Uncle Wallace is like her in so many ways I wonder if all this airing of the Goodridge past might have triggered his sudden need for the isolation and anonymity of city life. His need to get away.

Uncle David and I walk again, down to the wharf, up to the Mount. According to Steven Mills, the four cannons were actually moved to Renews from Ferryland long after the Dutch invasion, probably during the American Revolution. They had been in Ferryland since the 1670s and were quite useless by the time Renews got them a century later. The cannon balls were welcomed in the house, however; Uncle David remembers his mother using them as doorstops and bedwarmers, and some family authorities attest that they were also employed in the kitchen to mash vegetables into paste to make colcannon. Uncle David says my mother would roll them across the wooden floor of the attic to frighten her sister Sheila, who was terrified of thunder.

We walk north along the coast, following an old footpath below the Mount to Aggie Dinn's Cove. Aggie Dinn has been dead a long time and her house is no longer there, but Uncle David points out the clump of willow where it once stood. The cove is low and sloped, like a natural boat launch; the water slides up the bedrock to sniff at our feet as we pass. Not far beyond that Uncle David has built a small greenhouse, and we stop to inspect his tomato plants.

Some of the photographs that Uncle Wallace gave Steven Mills were of the Goodridge premises in Renews as they had looked in the 1930s. They show two huge warehouses built right out onto the old wharf, with a scattering of smaller, single-storeyed buildings and wharves around them. It was a prosperous-looking enterprise, the buildings painted white, the roofs neatly trimmed and shingled. Fish stages sloped up the hill, the fences around them straight and well maintained. A dozen schooners lay at anchor in the harbour, sails furled, their holds perhaps still smelling of sugar or tea. In one of the photos, two brand-new cars are parked in the yard in front of the premises; they look to me like gangster cars, as though Al Capone had dropped by to pay his respects. I ask Uncle David what happened to the business.

"Avalon, that was Father's brother, sold or gave his shares to his two sons, Harold and Norman. Harold ran the company, but he ran it badly. Ran it into the ground, in fact. He eventually sold it to a man named O'Brien, who did nothing with it until finally, in 1972, the buildings burned down." Uncle David speaks as though the fire had come as something of a relief.

Before leaving Renews, I walk down to the spot at the head of the wharf where the old house had stood. I have some digging of my own to do. Just before my mother died, she told me that she wanted some part of Newfoundland sprinkled on her grave, and I think nothing would be more appropriate than a handful of earth from the site of the house in which she had been raised. Looking about, for I did not want to be observed, I ducked under the willows, got down on my hands and knees, and scrabbled in the dry soil with my fingers. I felt foolish, a fifty-five-year-old man on his knees in the shrubbery, digging with his forepaws, filling a bag with dirt. But I also felt a sensory connection with everything around me; these were my shrubs I was smelling, this was my dirt under my fingernails.

If history is a river, then it has to end somewhere. If we think of it as a sea, on the other hand, then it is always there, always part of the present, and I am tending toward the latter view. When Steven Mills brought me into the schoolhouse and showed me the artifacts he was unearthing from the tavern site, I experienced an almost electrical charge of immediacy. The shelves were arranged with things that could have been in use in my mother's day: wonderfully shaped wine bottles, thick-stemmed glasses, bags of broken pipes, bits of crockery. Steven could identify most of the makers. He had reproductions of all their marks, and drawings of seventeenth-century English alehouses – Robert Bird's shop in S. Lauren Lane, London, for example, where in 1631 fine ale was sold "at the sign of the Bible." In the drawing, men in wide-brimmed hats sit at a linen-covered table on which await trenchers and a plate of roasted fowl. Steven also had a portrait of Sir Henry Cary, Viscount of Falkland, who colonized Renews in 1629, shortly after Lord Baltimore withdrew his settlement from Ferryland. Ben Jonson wrote a careful Epigram to Cary, who was then also Lord Deputy of Ireland and a powerful figure in the court of James I. The poem ends, "He'st valiant'st that dares fight, and not for pay, / That virtuous is, when the reward's away." All of that history seems as much a part of present-day Renews as are my mother's memories, my uncles' familiar footsteps, my own small bag of dry earth.

My mother won't be going home again, but I will bring a bit of home back to pour on her grave. The Newfoundland earth will mix with the gravedirt over her ashes and, eventually, with her. Merilyn will plant a few flowers and their roots will aid in the mixing. As I stand up I will hear the cry of a baccalieu bird out on the harbour, and Uncle David will tell me that coyotes have come to the island from Labrador, floating across the strait on ice pans in winter, and how they can sometimes hear them howling in Renews from their den-sites across the Irish Loop. They howl like sailors in a tavern, he will say, the sound of their voices carrying eerily over the water. In baseball, I will tell him, the need to be safe at home is too strong to be constrained by time.

ACKNOWLEDGEMENTS

Writing essays is like going for a series of walks in unfamiliar territory; no two walks are the same, and one always runs into someone just when one needs advice or direction or just to pass the time. I have been generously helped along on these journeys; most of the people who walked along with me are mentioned in the essays themselves, but some are not, and some of those who are deserve to be mentioned again.

Zal Yanofsky told me to look up Ed Good when I was in Vancouver, and Ed accompanied me and provided valuable backup on my search for the last Crested mynahs. Charles Wilkins conducted a peripatetic e-mail correspondence with me as he walked across northern Ontario, and that was before wireless. Graeme Gibson put me on to Dave Mossop before I left for the Yukon, and Dave and his wife, Grace, helped make my four-month stay there both pleasant and exciting. So did Erling Friis-Baastad, Patricia Robertson, and Leona Etmanski. Misty MacDuffee of the Raincoast Conservation Society spoke to me passionately about grizzlies one rainy afternoon in Sidney, B.C., and it was Tove Reece, of

Edmonton's Voice for Animals, who sent me a videotape of coyotes being tormented in Alberta. And David and Wallace Goodridge gave generously of their time, friendship, and memories during my stay in Renews, Newfoundland, as did Stephen Mills of Memorial University.

Many thanks, of course, to James Little, who saw these essays as columns at *explore* magazine; and to Alex Schultz, who saw the columns as essays for McClelland & Stewart. Jenny Bradshaw helped smooth these essays together and saved me from many embarrassments. Bella Pomer has been my second reader for many happy years. And I am deeply grateful, as always, to my wife, Merilyn Simonds, my first reader, for her wisdom, enthusiasm, and inspiration.